THE BITE-SIZED WIN MINDSET

Modernizing your software delivery process in little ways

DEREK S. TORRENCE

Derek S. Torrence

Printed Worldwide
First Printing 2023
First Edition 2023

ISBN 979-8-9881136-1-4 (eBook)
ISBN 979-8-9881136-0-7 (Paperback)

Library of Congress Control Number
2023906548

For my parents, George and Vicki

The readers of this book now know who to blame when they come across any bad jokes or sarcastic humor.

THE BITE-SIZED WIN MINDSET

Contents

Introduction

I have been a software developer since 2006. In my time, I have been involved in and seen many iterations of the software development lifecycle. I have worked with very small teams (roughly five developers and QA agents) as well as massive teams that work across several global time zones. Throughout most of my tenure, I have been in leadership roles within the teams that I was a member of. While my bread-and-butter has been software development, my passion has been finding ways to help my teams succeed by improving the software development and delivery process.

Early on, I found that assigned tasks were easier to complete when the requested task was small. I associate the desire for smaller tasks with my time in college, when I excelled at working on homework assignments that were iteratively developed as the semester went on, as opposed to singular end-of-semester projects. With this insight, I decided to break the larger tasks assigned to me down into several bite-sized tasks. I always have a notebook with me, so I would just jot down a checklist of items that had to be accomplished in order to complete the full task. Once I had my work broken into smaller tasks, I could figure out more reliable time estimations and plan better for a smooth delivery.

Sometime around the year 2016, I met my first Agile coach, Tammy. She quickly had an impact on our organization, as she was able to quickly recognize that our financial software was a complex monolith of spaghetti code and helped us organize our team in such a way that, even if we did not have a clear understanding of how the offending section of code worked and how it could be fixed, we could formulate a plan that would allow us to deliver our software in reasonable chunks. In turn, we could keep everyone invested in the process happy. This is the effect of what I now call the bite-sized win mindset.

I was speaking with Tammy recently and told her about this book. I thanked her for her influence on my career and we reflected on some of our past experiences with writing bite-sized stories. Her remark to me was: "I like to say it is about the flow of work through the delivery system." That is probably the best way to describe our goals for changing our processes to accept more of a bite-sized win mindset. Many teams need a way to more effectively manage their work, and converting to bite-sized stories will help control that workflow throughout the delivery process.

Since working with Tammy, I have studied and experimented with this bite-sized approach to software development further and expanded upon what I learned from my time working alongside her. I have managed to help my teams transform their belief that it is acceptable for software to take months to develop and to, instead, think of software delivery as a way of providing small bits of code that progress the full project while keeping customers happy and allowing your team the full time to finish development. Make no mistake, if you break work down into smaller chunks, you are not instantly guaranteed a faster delivery, but delivering work more frequently in this way will buy your team the full time required to finish the software. The end goal of a software team is to deliver high-quality code without jeopardizing timelines that have been agreed upon for delivery. I feel strongly about the bite-sized win mindset and finding bite-sized wins wherever possible. This is a topic that I have been passionate about for a very long time.

Throughout this book, you will find different ideas, personal stories and hypothetical use-cases for the bite-sized win mindset. This book is primarily written with the assumption that you are in a decision-making role within a software development team. From time to time, I will refer to you as being within a certain role on your team, but you could be in any role that leads your current team. Do not worry if you are not in a role like this in your current job; just use the old noggin to imagine with us! Anyone involved in the software development process will gain some insight and benefit from the ideas and advice provided in this book.

In case it was not obvious, I am no artist, but I do hope you enjoy my attempts at creating something artistic and humorous for you to enjoy as you browse through the pages of this book. Just in case I was not clear, here is an example of my art:

In this book, I will begin by outlining what "bite-sized win" really means and how it can apply not just to software development, but how it can help organize your life in general. In Chapter 2, I discuss how to let your team grow by facilitating self-organization, where your software developers, as well as your product owner, can identify work that needs to be tackled and how you can work together to agree upon ways to handle the software delivery process to adhere to a timeline that will satisfy your clients. Chapter 3 will provide practical advice on how to divvy up tickets in a bite-sized win fashion. Chapter 4, The Toss-5 Method, will provide a method to help to train your brain into doing a few small things every day.

Chapter 5 will help you think about who is involved in your software development process and how to address concerns that they may have when adopting a bite-sized approach. Chapter 6, Separation of Concerns, describes a concept known to many software development professionals and explains how it relates to the bite-sized win mindset. Chapter 7 discusses how failures can be used to your advantage, and advises on how to encourage a cohesive team mindset along the lines of: "not only do we win together, but we also fail together." Chapter 8 will help you find a way to increase company awareness of your team's wins and advise on how to (tastefully) inform members of your organization of your successes so they can have confidence in your team's abilities. In Chapter 9, we will recap on the major aims and lessons of the book and the bite-sized win mindset.

THE END GOAL OF A SOFTWARE TEAM IS TO DELIVER HIGH-QUALITY CODE **WITHOUT JEOPARDIZING TIMELINES THAT HAVE BEEN AGREED UPON FOR DELIVERY**

Chapter 1 | What is a Bite-Sized Win?

You may be wondering what a bite-sized win even is: it is the completion of one smaller part of a much larger task, and is a way of providing constant positive reinforcement to everyone involved in your software development process, be it your software developers, the business or the client. This concept is not exclusive to software development or any specific industry, but is a general concept that individuals can and do use in their day-to-day lives. I have seen many different approaches to providing bite-sized wins, regardless of what its users call it or to what area it is applied. I am sure you have also seen numerous articles online that discuss ways to break down your duties into smaller tasks, which encompass interests as widespread as organizing your home and studying for an exam. Whatever the topic, the method is always the same: when you take on a task that proves to be a large undertaking, you should not aim to complete the work in a single sitting, but rather break it up into smaller chunks. At the end of the day, when you see progress by completing these smaller tasks, you can better appreciate the work that you are doing.

Most people see a task as a body of work that needs to be completed all at once; I want you to start thinking of a task as a small bit of work that can be done as part of an incremental delivery. When I refer to an incremental delivery, I am referring to the concept of each delivery building a little more functionality on top of the deliveries prior. You start by identifying a large goal (either a task or your project) and you use several software deliveries to build up to delivering your goal in whole. Each delivery should stand on its own as something tangible that does not adversely affect your system's daily operations.

Gains of subtasking

As previously mentioned, my professional experience is in software development. Throughout my career, I have noticed this

pattern: the business will deliver a set of requirements to the team, which includes some desired input and some desired output. During the software development lifecycle (SDLC), the team will focus on getting the requested work done but will neglect to keep the business informed of statuses between

deliveries. Instead of considering how to keep the business happy, the team will often focus only on how to keep the business at bay while the software development process is underway. I believe that the ideas of business happiness and software developer happiness go together. The team therefore needs to provide feedback to the business while simultaneously requesting that the business respects the amount of time that software takes to develop. One way to encourage this harmonious working relationship is to incorporate the bite-sized approach to subtasking.

For instance, if you have ever worked on a software development team, you can empathize with the fact that code will always become outdated, but refactoring for performance is usually not accounted for in the time allocated to the team. When your team breaks work down into smaller tasks (or subtasks), the team will be given a newfound amount of time allocation for any high-priority, low-prioritized legacy upkeep that has been neglected over the years. The reason for this is that smaller tasks will result in the ability to squeeze in small units of work at the end of the development cycle. There will inevitably be time before a project's delivery date when one or several software developers have completed their work. If the software developer faces a large refactoring task, they may not have time to complete it, but if the task has been broken down into smaller, bite-sized requirements, they have a better chance to tackle at least some of the requests.

It is actually possible that your team is already working with the bite-sized win approach but has not realized it. After your team has completed scoping the work for the upcoming sprint, what is the next usual step? Do you send your team back to their desks and wish them good luck? My guess is that, in most instances, you first ask the software developers to create subtasks for their tickets. For those readers who are not from a software development background, a ticket (also referred to as a task) describes the work being requested, in most cases by the client. Subtasks are a way for a software developer to visualize the work that is in front of them; they also help the Scrum Master, development lead, or anyone else in a decision-making role on the team get a better understanding of where the software developer is in the software development cycle for a requested task.

I have just mentioned the term "sprint." This is something that we will explore often in this book, so now is a good time to explain it: a sprint is a set time period in which the team is expected to complete a set of work that has been agreed upon. This timeframe will often be two weeks, but a team is free to specify any amount of time. Being able to consistently complete allocated work within the sprint is what the team hopes to achieve. However, at its core, the idea of the sprint is that the team has agreed that they will commit to blocks of work to be completed on a regular cadence. The amount of work completed is dependent upon how well the requirements and tickets have been written. Within the sprint, there may be a lot of tickets or just a few, but the goal is to deliver all the work that has been agreed upon by the end of the sprint.

Potential pitfalls of subtasking

Subtasks are a great organizational tool, but only when managed properly. You will often find that the software developer does not keep the status of their subtasks updated. Instead, the software developer will often have a list of subtasks that they briefly reference in the morning so they know where they should focus their energy for the day, and then they will neglect to update the status of those subtasks for the remainder of the project. This does not satisfy the decision-makers and it means that the business does not see the value in allocating time to subtasking. Failure to update statuses can also negatively impact the client's perception of your team's ability to maintain transparency during the software development and delivery process, since the tasks will be inadequate in providing enough real-time information about where the team is

at in the software development process without scheduling time for a verbal update.

In a previous role of mine, my team's leadership tried to force the team to create subtasks before giving tickets points in our backlog grooming sessions. We would sit together, discuss the amount of work that was required to deliver the task and then ask the group to rattle off some subtasks that could be used as a guide for the software developer. The positive effect was that we could help our less-skilled software developers enhance their skills and overall understanding of our system by assigning them progressively advanced tasks, but the drawback was that nobody updated the statuses of these subtasks, and the business viewed this method as a waste of resources, time and money. When using subtasks or any other method to track work, remember that the business requires tangible metrics in order to justify a change in process.

When subtask updates are neglected by the software developers, the team has not been successful in addressing business concerns over the lack of transparency in the software development process. Keep in mind that the business should not have to refer to a list of subtasks just to find out what the software developer did that week, especially since in some ticketing software, isolating subtasks for filtering, sorting or reporting is often a difficult goal to achieve. It is the tendency of software developers to treat subtasks like minor notes, so most of the subtask titles are not very descriptive and would not serve the purpose of providing clarity about what work has been done once the subtask is completed.

Instead, I would suggest that in your ticket-tracking software, you create units of work as a task-type ticket, instead of a subtask-type ticket. Of course, every software is different, but in most of the major software available, a task ticket is meant to be used to define a deliverable, whereas a subtask ticket is meant to give the software developer guidance on what work needs to be done in order to accomplish the task. When asking myself if a unit of work should be a task or subtask, I ask what the result of the ticket will be upon delivery. If the subtask can stand on its own and contribute something to either the user interface or the backend codebase, I move that into a task (or story) and allow it to move into the main repository. Some teams will break up tasks into front-end and backend, but this approach is personal preference. It is possible to encompass all tiers of the architecture and still have a task that is small enough to be considered a bite-sized win. We will discuss determining appropriate sizes for tickets in Chapter 3: Tasking for the Bite, as well as Chapter 6: Separation of Concerns.

Reworking the workflow

When guiding your team towards incorporating bite-sized wins into their working cadence, determining *how* work will be presented
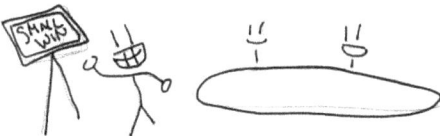
is just the first step. After the team is comfortable with the quality and content of the tasks being presented, your next step will be to modernize your workflow to allow these tasks to comfortably exist when your team plans for future work. I would like to present two different workflows: legacy and modern. The legacy workflow is more of a

traditional workflow, which I have experienced numerous times throughout my career; it does not encourage roles to work together during the software delivery process. You may notice that the modern workflow contains more steps, but it encourages the team to work together through each step, regardless of their role.

The legacy workflow

Consider the following workflow:

1. The client makes a request.
2. The product owner accepts the request.
3. The product owner translates the request into a software developer-friendly task.
4. The task is sent to the software developer.
5. The team awaits word about the completion of the task from the software developer.
6. If issues arise, the software developer works with the product owner or other software developers to resolve them.
7. The client receives a software update with the requested work.

The workflow described above is akin to a Scrum team's workflow, but it violates one major rule: nobody is working together until a problem arises. A methodology like Scrum hopes to mitigate the mask of uncertainty by involving the product owner in the daily process, who in turn will keep the business informed about the status of the work assignments. However, even with a product owner being directly involved, there may still be requirements that are misunderstood or overlooked until the final delivery has completed. The unfortunate result of this workflow is that the client and product owner are both upset about the amount of time that it took to deliver the software, and the software developer, who faced unforeseen delays due to the team's misunderstanding of the issue, is now viewed as the reason why the task took so long to deliver, when really the fault is shared among everyone involved and lies in the amount of time that it took for the team to return to the client and product owner for clarification. This workflow potentially becomes an endless chain of blame, until the blame is ultimately set on the software developer, who was not involved in the planning phase and could not weigh in on the amount of effort that would be required to complete the task.

The reality is that the client can make requests all day (and they will, if left untethered). This is fine. This is how software development works. But the product owner, along with the software developers, should take ownership for and help prevent excessive time requirements for task delivery. The mindset of the software developers and product owner should be that any task should be properly groomed before it is presented to the team for acceptance into the current or upcoming sprint. A properly groomed ticket is one that will provide the software developer with a clear understanding of the requested work, such as a full list of requirements, any relevant documents attached, and it should clearly define the problem as well as the expected outcome when the work is complete.

The criterion for a properly groomed ticket comes down to what the team agrees upon in their initial team delivery agreement. The team's delivery agreement is a set of agreed-upon rules which dictate how the team will function throughout the software development lifecycle; these rules can be altered at any time. One of the rules that I like to impress upon my team is that the client, as well as the business, should never be unsure of the status of a project at any moment. This transparency should not come from the team, but rather from the work that is delivered. In order to keep the client and the business happy, in the team's delivery agreement, we must agree upon a consistent cadence of delivery, regardless of size, in order to create the perception of transparency to avoid being interrogated for status updates on a regular basis. We will dive deeper into the team delivery agreement and how to incorporate the bite-sized win mindset in Chapter 2: How Teams Self-Organize.

Within the corporate software-feature-request totem pole, the software developer is on the bottom. It is not appropriate for the software developer to take the heat for decisions made above them. The team should aim to positively impact the client's experience while ensuring that the team (as opposed to a single team member) is prepared to take on any blame for software delivery issues from the client and/or the business. The team's delivery agreement is a way of acknowledging that the entire team holds responsibility for keeping the project's stakeholders informed throughout the delivery process, and no member of the team should be singled out as being responsible for delays in the process. Information should be provided to the client and business with the goal of full transparency and should focus on how successfully the team delivers reliable, stable software on time. If there is ever a moment when a single team member stands out, whether positively or negatively, it means the team is not adhering to their delivery agreement and changes to the working atmosphere should be addressed immediately.

Requests for work can come from anybody within the organization or directly from the client, so the team's ability to deliver should not be focused on who made the request. If a request is handled differently based on who made it, the team may approach the delivery with a skewed set of rules. For example, if the request came from your CEO, you may consider it a high-priority, quick delivery event, and therefore create tasks with less information. This may result in software with many bugs that could have been prevented if more information was provided in the task before it was moved into the sprint. This is why it is important to have a single set of rules for all tasks, no matter where the request originated. Context is very important to a successful delivery, but the requesting party should be irrelevant. Instead, the team should strive to put their best effort forward, regardless of who may be evaluating their work. Think of work as an assembly line, where a machine sets a box on the conveyor belt, the team attaches requirements to the box, and then the box continues down the line and receives various bits of work from software developers until the box is complete. The box is then removed from the conveyor belt and handed to another machine that will deliver the box. An assembly line does not care where a box comes from. An assembly line does not care where a box is delivered. The assembly line will continually package and deliver boxes until the final whistle blows. Strive to be an assembly line, but have fun while doing it. And, as we all know, assembly lines work bit-by-bit, putting together subtasks to accomplish a whole. Let's see how we might incorporate this thinking into an improved workflow.

Modernizing your workflow

1. The client makes a request.
2. The product owner accepts the request.
3. The product owner creates an initial set of requirements to present to the team.
4. The software developers work together to decide how many tasks to break the request into.
5. The product owner and software developers decide which features are the *most important* and map out delivery stages.
6. Each task is given clear instructions on how it should be completed.
7. Each task is given clear instructions on which tasks have dependencies.
8. Each task is assigned to a software developer (not necessarily the same developer for all related tasks).
9. Every time one of the small tasks is released, the client reviews the unit of work and provides feedback.

When questioning the best way to approach your workflow, you should ask how you can get the most effort out of your team members without burning them out. The modernized workflow above focuses on including your entire team every step of the way. In previous workflows, the product owner (arguably, the business analyst) would take ownership of writing out the full requirements for the request and would hand over a mind-numbingly long document to the software development team, likely avoiding engaging the team until the deadline approached and the client asked for an update. This may be a little exaggerated, but should shine some light on why processes require the kind of modernization outlined above. Consider how quickly technology evolves; why should the software development process be hindered and not evolve as quickly?

I would like to propose a hypothetical situation: you work on a team developing a web-based product. One day, out of the blue, the client asks for a new feature that will enable their website visitors to customize their website experience. Visitors should have a control panel with several settings that will fundamentally alter the look and feel of the website. As an added benefit, the client wants to provide a minimum of ten themes the visitor can use as a template to help them learn what can be customized. This work will undoubtedly take several months to fully deliver. The team's product owner delivers a hefty, 200-page document with everything that you would ever need to know about the client's request. How should your team approach this?

Simple: work with the product owner on a solution that will give the client some bite-sized deliverables over the next few months. Hold several planning sessions that will break down the 200-page document into bite-sized pieces, which enable the software developers to comprehend every aspect of what is being requested. By the end of these sessions, all members of the software development team should understand the reason for the request and the desire for the delivery and should agree upon a schedule of delivery that provides bite-sized wins within each milestone.

Perhaps the first step is to create the page where this feature is going to live. You might put up an under-construction message. Then, as software developers add content, this page can be populated. Focus on creating tasks related to the requirements document which can guarantee an iterative delivery, and subtly build upon previous requirements. In summary, start with an "Under Construction" page, then add a few elements that show an outline, add a little logic, sprinkle some magic on top and reap the benefits of a happy client.

The biggest benefit to this approach is that not only are you delivering something the client can preview and use before final delivery, but you are also giving the client the ability to refer to that page frequently and gain valuable insight into the progress. If required, the client could intervene and request changes as development progresses. It is undesirable to deliver a months-long project to a client and watch the smile fade from their face as they inform you that they asked for a red ball, but you gave them a blue pyramid. The blue pyramid sure does look fancy and that ombre fade effect would make even the strongest of men weep at its beauty, but the client intended to use the red ball as a clown's nose; a clown with a blue pyramid for a nose is not silly enough.

The practice of holding on to all work until a feature has been fully completed frustrates everyone if something goes wrong with requirements interpretation. Should a feature take *x-months* to produce, the client will be unable to review the complete feature for *x-months* and so it is *x-months* until the client discovers that the team did not properly interpret the initial requirements. By taking the time to break down requirements into bite-sized tasks, a plan for iterative releases can be integrated into the delivery schedule and the client can receive constant updates.

The initial planning stage for a delivery schedule is the time to decide how granular your tasks will be. If a timeline is tight and there are not many releases until the feature is due, then you would likely plan for small tasks that can be worked on simultaneously by software developers. On the inverse, if a timeline is a bit more liberal and forgiving, then you can break tasks down small enough that it does not matter if only one software developer is tackling each task; the emphasis will be placed on which pieces to deliver to the client so that the team can better prepare for incoming issues or changes to requirements.

Summary

What is a bite-sized win? Completing one smaller part of a much larger task, and a way of providing constant positive reinforcement to everyone involved in your process, be it your software developers, the business or the client. Bite-sized wins can be achieved by ensuring your team's tasks are appropriately sized, but they may require your team's workflow to be modernized. The team must work together as a single unit and accept responsibility for all aspects of the software delivery process.

One advantage of the bite-sized win in software development is that the team can ensure that the client is able to look at what is being developed while it is being developed, ensure they are happy with the progress that is happening and intervene at any point if required. Another is that work can be conducted simultaneously with multiple developers in order to satisfy a tight deadline. Proper planning can identify and tackle the issues at hand. If the team initially plans for an iterative release for a feature, it is still possible that the team can take those tasks and assign them to multiple software developers for a quicker release.

THE REALITY IS
THAT THE
CLIENT CAN
MAKE REQUESTS
ALL DAY
(AND THEY
WILL, IF LEFT
UNTETHERED)

Chapter 2 | How Teams Self-Organize

As a leader, you will work with your team on organizing work and delivery. At some point, an effective leader will find a way to separate themself from that process and allow the team to self-organize. In this chapter, let's explore your role as a leader who is responsible for organizing the team and help you to evolve into a *servant leader*, who instead allows the team to self-organize. A servant leader does not give orders to the team, but rather accepts direction from the team when required to remove roadblocks. When exploring the bite-sized win mindset, teams should be willing and able to think of their work in smaller units without requiring a leader to tell them what those smaller units should be. The practice of self-organizing will enable the team to decide how much work fits their capacity and gives them the power to redefine their capacity for work as the team expands, shrinks and evolves.

Why should your team self-organize?

For anyone who has worked in a software development environment, the concept of organizing teams is nothing foreign. Environments can be hectic if nobody is actively working to achieve some semblance of organization. Think about your team members; is everyone just taking tasks and producing work? The more likely scenario is that, at the beginning of every sprint, one person gathers tasks and assigns them to software developers. After the tasks have been spread across the team, there are probably daily stand-ups in which someone asks

about the status of assigned tasks and hopes to get an ETA from each member, which they can relay back to the client. Then, at the end of the software development cycle, the team comes together and discusses what was accomplished, as well as what kind of lessons can be learned from the experience.

However, relying on a single member of the team to organize everyone involved in the process is quite a large expectation and detracts from other work. There are only so many hours in a day and unless that team member was hired with the sole purpose of organizing the entire team, they likely have other tasks and duties that require their attention. It is therefore important to explore how the software development team can self-organize. *Self-organizing* is when the team can be left unmanaged (read: mostly unmanaged) and is still able to cooperate to accomplish the requested workload. A self-organizing team does not require a defined leader in order to function. Instead, the team makes decisions as a single unit and has a democratic way of agreeing on their direction.

In the world of Agile software development, most teams strive for the autonomy of self-organization, and will seek to apply a set template for a swift path to completion. Some teams will formalize organization with a methodology like Scrum, Lean or SAFe. Most teams have a homebrewed flavor of one of these methodologies. Others will not have anything formal and will just commit work until they are ready to ship. In most of these cases, the methodology (or lack thereof) is decided by someone in a leadership role within the department. Once the foundation has been set, the team will be able to explore various techniques to find the one that works for everyone involved.

A leader, an owner, and a master

When a team is self-organizing, certain roles may seem to make sense as leaders. Product owners, for example, are responsible for gathering requirements, but expecting them to lead the team is not realistic because their focus is on gathering requirements and working directly with the client on their requests. A Scrum Master would seem like the best fit, since their role is to (by definition) be the master overseer of your Scrum team's implementation and execution, but they also do not fit the mold of an appropriate team leader because they serve as the team's shield when trying to finish their work in the sprint, as opposed to directing the team on how to finish their work. Let's explore these roles and how they will fit in with your team's goals for self-organization.

The self-appointed leadership

It is natural for your team to desire some form of leadership within the ranks. Without a leader, a team may question who their voice is. However, once the team learns how to operate as a unified voice, self-organizing with no designated leader becomes an easier process to imagine and follow. This said, even in self-organizing teams, the reality is that some form of leadership will rise. Some team members are naturally louder than others, so more timid members may allow them to rise to decision-making. I would not suggest trying to suppress that leadership when this happens. If the team decides that having one member serve as their voice is best for them, then that is how they are going to operate. It is fine to give the team a friendly reminder on occasion to help them understand that everyone has an equal voice, but the team dynamic will naturally mutate over time.

Some of the most successful teams that I have been part of do not have a designated leader. Instead, the team works as a single unit that discusses ideas as they are presented. The team decides whether the idea was good, great, or awful. When the team is self-organizing in this way, there is no need for a manager or software development lead to sit at the helm to handle day-to-day organization. The team can identify the work that is required, then formulate a plan for how the work will be completed by the assigned deadline.

The product owner

Workload management is where teams will struggle the most when attempting to self-organize. This struggle is why in recent years the product owner role has become part of team structures. The role of the product owner is to be the liaison between the client and the software developers, responsible for requirements gathering and grooming without serving as an extension of the business in terms of setting schedules for delivery.

In the past, the leadership at the corporate level would hire a business analyst who was tasked with speaking with clients on behalf of the organization and writing the requirements, then throwing those requirements over the fence to the software developers and cackling maniacally while watching them duel to the death over who was allowed to work on each shiny new feature. As time went on, it became apparent that isolating this requirements-gathering process from the core software development team was causing delays in delivery, extended signoff times and an overall break down in the confidence in the software development team. It became clear that it is important for the team to include all members of the organization who are responsible for the work. Businesses therefore introduced the idea of the product owner [cue the superhero music...], whose role is to create cohesion between the software developers and the project requirements. However, the product owner should be part of the team, and not be seen as dictating the work that will be accepted by the software development team.

How does *the product owner* differ from *the business*? We have used these two phrases frequently and there seems to be a lot of overlap, so I will provide some clarity. It may be helpful to look at "product owner" as a role (or career) and "business" as a label. Let's take this a step further and separate the product owner from the business analyst as well, since the product owner is part of the software development team, and the business analyst is technically part of the business. We will assume that your organization no longer has a business analyst role and has moved towards the product owner role instead. When we talk about "the business," we are talking about all members of the organization who are not directly involved in the software development process. The product owner is the role within the organization, which can be held by one or more individuals, whose purpose is to gather requirements and hold ownership of them all the way through the delivery of the work, as opposed to just identifying areas that require improvement and requesting someone else address them.

The Scrum Master

In self-organized teams, it can be easy to see the Scrum Master as the closest thing to a leader (or, more appropriately, a decisionmaker), but really, they serve more as a barrier between the development team and the business. The Scrum Master will schedule meetings to discuss the work ahead and help the team to identify pitfalls that could impact a deadline. If a Scrum Master is serving as a direct leader within the team, you may need a new Scrum Master.

My favorite Scrum Master of all time, Tammy, served as a warrior with two swords and a shield. She would help everyone to keep track of their time, eliminate unnecessary meetings, coordinate pushing deadlines and make sure the team was always aware of pending work when members felt like they had nothing to do. In order to prevent the team as seeing her as the leader, she would make sure that she facilitated discussions, instead of leading conversations. Tammy would schedule meetings with defined topics and agendas, then leave the discussions up to us. She was notorious for ensuring our meetings stayed within time limits, and if our topic required further discussions, she would not allow us to continue the conversation after the end of the meeting; instead, she would either schedule another meeting or ask that we start an email thread with everyone who was interested. Without providing the solutions to our problems, she was able to have the team view her as an essential part of the team, as opposed to someone who was in control of the team.

A team coming together to self-organize is a very democratic process and is crucial in the world of software development. A team should be able to work cohesively without relying on their product owner or Scrum Master in order to make a move. While the product owner is responsible for initiating the requirements-gathering process with the client, the team is responsible for ensuring the quality and content of the tasks that are created from those requirements. The Scrum Master should be working in the shadows to put out the fires without serving as the leader in how the team handles their daily workload.

You may have noticed that I put heavy emphasis on Scrum throughout this chapter. The intention is not to preach the benefits of Scrum and convince your team to adopt it, but rather to explain how roles within a team could, if you are not careful, step on each other as you attempt to encourage self-organization. Scrum just happens to be a popular Agile paradigm and I feel that most teams can relate to it. I would encourage your team to see which methodology works best for them. With that said, I would like to give a quick introduction to Scrum so that we are on the same page when I present some of these terminologies.

A quick introduction to Scrum

According to Scrum.org:

Scrum helps people and teams deliver value incrementally in a collaborative manner. If you are just getting started, think of it as a way to get work done as a team in small pieces at a time, with experimentation and feedback loops along the way.[1]

Scrum is an Agile methodology that helps the team to organize in such a way that every member of the team is involved in all areas of the software development lifecycle. It therefore fits perfectly with the idea of self-organization. Every member of the Scrum team has their own role, yet everybody works together as a single unit to organize work and deliver projects.

Traditional team workflows vs Scrum

In a traditional software team, your team would not have a product owner. Instead, your team would have a business analyst, whose role (as we just discussed) is to gather requirements and simply "toss them over the fence" to let the software developers

---

[1] https://www.scrum.org/

manage the work. The business analyst does not care to hear back from the software developers until the request has been properly fulfilled. In that traditional workflow, instead of a Scrum Master, you would have a project manager (not to be confused with a product owner), who would decide how long it should take to deliver software to the client.

With the onset of Agile methodologies, this has become an antiquated way of approaching software development. One major issue with this approach is that the team will often have obstacles along the way, be they related to comprehension of the requirements or conflicts with existing business logic that had been previously requested. To resolve these issues, they need to consult someone from the business analyst team, but in this system, they are unable to because the business analyst has already moved on to their next task. To combat this, Scrum utilizes a product owner to represent the business and client and a Scrum Master to corral the team. Scrum requires the presence of both roles, just as much as the software developers from the moment the request is made until the final delivery.

Another issue with the traditional software team workflow is that business analysts are not interested in collaborating with the team on how to deliver the final requirements, which is discouraging to other team members and often results in the team neglecting to consult with the business analyst when issues arise, since they do not see the business analyst as concerned with finding solutions or caring much about the process in general. Everyone in the process ought to have a sense of pride in delivering work, including the client, who should take ownership for the implementation. When using the bite-sized win mindset, the team receives constant reassurance that their efforts have not gone unnoticed. This ensures that everyone on the team gets those endorphin boosts on a regular basis, and this encourages them in their work. If a member of the team does not see value in taking pride in their work, then they will not contribute to the best of their abilities, as is often the case with the old mindset of the business analyst. The business analyst is happy that they created requirements out of a few conversations, and they will be happy again when they can tell the client that the work has been completed, but is uninterested in the steps in between; I find it mind-boggling when the business analyst does not appreciate the efficiency of the process, or feed off the team's energy to find new ways of accomplishing the work requested.

The bite-sized win mindset works perfectly in the world of Scrum because Scrum discourages large workloads being allocated to single team members. Scrum wants to help the team to organize in a way that allows everyone to support each other. The bite-sized win mindset aims to help the team to break down the work into even smaller units of work and deliver smaller bits more often. A big benefit of combining Scrum and the bite-sized win mindset is that, due to your team's current Agile implementation, your team may already be used to delivery schedules and will be open to the idea of a greater frequency of delivery. These combined concepts will encourage a team to work together on smaller units of work and ultimately shift towards a more frequent cadence of delivery.

Agility

One of the major goals of Scrum is to achieve "true agility" within teams. In order to be "truly agile," your team requires a mindset that allows changes to be easily implemented throughout the project. It is very common for a set of requirements to expand or shrink throughout the process, due to factors such as conflicting logic, the introduction of new information or even something as simple as a change in budget. In my experience, Scrum achieves this high level of agility by involving all members of the team from the initial requirements-gathering stage, allowing for more diversity in mindsets and more insight into how tasks will be handled. The benefit then continues throughout the software development lifecycle by ensuring that more members of the organization are involved when there are changes to requirements or business priorities.

However, it is not enough to just have everyone sit in a room; the team must be encouraged to provide their insight and expertise. If there is a glaring issue with the requested work, your product owner may not be able to identify it, but by getting input from everyone, you increase that chance that someone will identify it. Experience comes in the form of in-house application knowledge, as well as career expertise. Some software developers may not be as loud and opinionated about the process, but if they speak up, I suggest you listen. Requirements can take on a whole new life just because of one comment. Teamwork is very important to the core of a Scrum team.

In the spirit of "true agility," Scrum can be shaped into whatever works best for your team. There is no one-size-fits-all approach to working with the methodology and you will often make small changes for each different team. My previous Scrum experiences have been those of hybrid implementations that were tailored specifically for the organizations that were implementing it. When the team initially started to form, the organization would put some textbook rules in place that the team would try their best to adhere to. Over time, the team would keep what worked and dispose of any parts of the process that did not work for them. New ideas would form, and the team would shift focus again. This is a very common practice within Scrum teams, and it seems as though nobody really *does Scrum* in the absolute way that it is "supposed" to be done. Instead, the team will mutate over time and realize that a tailored approach to the team's organization can better help deliver work by the end of the sprint.

Team delivery agreement

One practice of Scrum teams is the creation of a *team delivery agreement* (or, in some cases, a *team working agreement*). The purpose of this agreement is to outline how a team will operate during their sprint. The agreement will specify a few criteria, such as the minimum amount of technical and non-technical information that should be required to be written in the task, maximum amount of time a task should take to complete, number of unit tests that should be written and even how many rounds of testing the quality assurance (QA) team should apply. I like to have my teams focus on the quality of their tasks to ensure that all members of the team, regardless of skill level, can do the work. The idea is to include everybody in the conversation and add or remove items from the agreement as necessary. The team's delivery agreement is not set in stone, and it is not uncommon for teams to refer to the agreement and update, add or remove items as the team evolves.

Not everyone uses Scrum

Back in 2012, I started working at an organization that was operating with more of a Kanban-type of methodology. With Kanban, tickets were generated during the requirements-gathering phase and moved forward until the body of work was sent to QA. If QA uncovered a bug, they would create a new bug ticket and assign it to the software developer, and then the software developer would validate the bug, fix it and move the ticket down the line once again. When the bug was resolved, the main work ticket, as well as any related bug tickets, would be moved ahead to the final stage and await final deployment. Work only moved forward, and it was discouraged to send tickets in the opposite direction. One of the benefits of this system was that we could gauge how often bugs were uncovered by QA. The drawback was that QA would often uncover bugs that were unrelated to the task that they were asked to test, and it would cause incorrect reporting on defects that were introduced during the development cycle. My development lead and I would constantly scrub tickets to ensure that we did not have incorrect bug-association reporting that could lead to incorrect reporting to the business. Kanban is something that Scrum sought to replace. (There is no reason why your team would be unable to implement a hybrid version of Scrum and Kanban, but if you are a

Scrum-absolutist, you are probably going to snub your nose at the very thought of *Scrumban* because there are a lot of key differences.)

Recognizing when to self-organize

There will come a point at which your team is mature enough to self-organize and you, as the current designated leader for your team, need to be ready to allow it to happen. Sometimes it happens organically on its own. Other times, self-organization needs a little push from you. You need to be able to recognize when it is time to encourage them to *begin* self-organizing. Consider the mother bird who allows her babies to take flight for the first time. If the mother bird were to discourage the baby bird from stretching its feathers, she will be forced to care for her babies for the remainder of her life. Allow your team to stretch their wings and take flight.

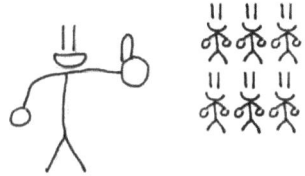

In order to support a team who is starting to self-organize, you must be able to identify when the team is ready. Often, this is when they start to realize that their method of gathering requirements allows them to work together more efficiently to achieve a successful delivery. Perhaps they have sat in on enough meetings that they now really understand the business needs. The team's maturity comes into play when they have a deep understanding of the business, as well as the software. This deep understanding will allow the team to make decisions that are not often left up to software developers, such as prioritizing tasks or outright rejecting requests based on lack of information. The team's product owner will be comfortable with speaking to the software developers on a highly detailed level and everyone will be confident that they are on the same page. Once the team can work in a cohesive manner, without guidance, they will appreciate the freedom that comes from self-organization. As a leader within your organization, you will reap the benefits of your time being returned to you.

The goal is to have a confident team that can take initial requirements, break them down into bite-sized chunks and then use those to deliver what the client is requesting. The team will likely have a few members who stand out and guide a lot of conversations. As the team continues to mature, there will be more voices in those kinds of conversations. Keep in mind that "the team" includes the product owner, so do not expect your software developers to self-organize in a bubble. If you notice that your team is excluding the product owner, make sure you intervene and guide everyone to work together. The software developers need to be able to speak with the product team, and the product team needs to have fruitful conversations with your clients and coordinate actions with the software developers.

One other key indicator that your team is already self-organizing is a reduction in questions for members of the organization who may be considered more authoritative leaders. The team will naturally gravitate towards anybody who tells them what to do and will often return to their old habit of looking to the external leader as someone who will make major decisions and take the fall when something fails. When the team is willing to make those difficult decisions without consulting more external authoritative figures, and the team is confident enough to fail, those leaders who were tied up in making major decisions are now able to focus their energy on providing the team with more resources, which will help everyone in the long-run. It goes without saying that nobody wants extra work. Unfortunately, leaders themselves will rarely realize that being the final word on team decisions is a level of work that never had to be done. The team should be willing and able to operate under the rules of the leadership within the organization without relying on the organization's leadership to dictate the rules daily.

As one example of being able to recognize business needs and make decisions accordingly, my old boss told me a story about how he was contracted to write software for a company that was, in many ways, so antiquated it was still using Excel spreadsheets for major business functions. They had a convoluted way of operating a particular process within their teams, so they asked him to create formal software to handle it for them. He instead found that the entire process could in fact be managed in a single Access database, without the need to build something from the ground up. He spoke with their managers about it and gave a few resources to explain how it was done, then left without creating the software. Sure, his boss was unhappy that the client opted to cancel their contract, and he would never admit to his boss the real reason why they did, but he was confident enough to take the honest approach: if you are overengineering your solution, your solution should not be honored. This holds true for management. If you have someone in a role that sounds more impressive than the tasks that they do, you do not need that kind of person in that role.

Leadership will remain at the corporate level

To alleviate any worries current leadership might have: the self-organizing team is not aiming to phase anyone out of a job. It is not stopping anyone from being a leader, and leadership is not neglecting the responsibilities of their role. On the contrary, the fact that the organization's leadership can lead the team to such a degree that they no longer need leadership means that they are leading the team properly, and the self-organized team should still include the higher-ups when necessary. There is some misconception in the business world that everyone should always be protecting their "job security." This trope is perpetuated by those with limited responsibilities. If a member of the organization can only prove their worth by counting how many people they are overseeing and not by evaluating *how* they are overseeing them, then it is possible that they should be replaced with someone who sees the value in a team that can operate without a visible leader. I do not mean to say that anyone deserves to be *automated out of a job,* but rather I believe that the right person for the job will always find something valuable to contribute.

When I was first learning about the concept of a self-organizing team, I was part of a team who thought we were already in charge of all decisions that had to be made during the software development lifecycle. It was not until our Agile coach separated herself from our discussions that we realized how much we were still relying on those above us to make sure we were taking care of our day-to-day operations. We learned about our lack of knowledge about self-organizing when we showed up to a requirements-gathering meeting and nobody was there except for the software developers and the product owner. After enough iterations of this, we found that our Agile coach and our development manager were not even required for these meetings anymore. This allowed everyone to get more work done.

Summary

Software development teams often share the same goal: deliver work as efficiently and reliably as possible. When a team is receiving orders from one person (such as a manager), the team will await their orders before making any decisions regarding how to designate tasks or how they should be prioritized. We should strive for a modernized workflow, wherein the team can make decisions without external leadership dictating goals. As we explored in this chapter, leadership can arise within your own team, but each member should be encouraged to share responsibility for any decisions made on behalf of the whole team. Your team may want to rely on their Scrum Master or product owner to represent the team, but that is not their role. We can use Scrum as an example of a methodology that attempts to encourage self-organization. However, know that you are not limited to Scrum, and there are plenty of homebrewed versions of different Agile methodologies to choose from.

THERE WILL COME
A POINT AT
WHICH YOUR
TEAM IS MATURE
ENOUGH TO SELF-
ORGANIZE
**YOU NEED TO BE
READY TO ALLOW
IT TO HAPPEN**

Chapter 3 | Tasking for the Bite

Before a team can start working, the team will need to create tickets. A *ticket* is a document that describes the work being requested, and it often gives information such as a due date and an estimation of the effort involved. In previous chapters, we spoke about tasks. While tasks and tickets share similarities, it is important to distinguish the *task* as the body of work being requested, and the *ticket* as the document that defines the task. In this chapter, we will explore ways to identify different types of tickets, how to recognize when a ticket is too large and some tricks your software developers can use to hide their work in the software until it is ready to be seen. When a ticket is properly defined, the task will be better understood.

Types of tickets

There are two categories of tickets that exist on a higher level than tasks and subtasks, which your team will likely create frequently: the epic and the story. The *epic* is the largest type of ticket and can be thought of as the scope for your project. An epic ticket includes all the information necessary to understand the final goal. Often, an epic ticket includes an attachment of the pages-long requirements document authored by the product owner.

Another ticket type is the *story*. Most teams consider a story to be the ticket for a task, which defines a smaller unit of work often based off an epic. A story should be large enough that it can deliver something tangible at the end of the sprint, while not being so large that the story may take more than one sprint to complete. As part of the team's delivery agreement, the team will dictate how large a story can be and will outline the minimum amount of information that should be included in the story for the team to accept the work. If the team finds that their delivery agreement definition for a story is incorrect, they can alter it at any time, but keep in mind that the change in rules could affect any stories that have already been created.

Note: most ticketing software will, by default, separate out stories and tasks, because some teams consider a task to be a unit of work that is either not part of a project or is an external unit of work that supports the story, such as defining and building the supporting architecture.

Analyzing the required effort

In the Scrum methodology in particular, tickets often contain a *point*, which is a number that tells you roughly how much effort is involved in the request. This number could be represented in different ways, and it is widely accepted that each team will likely have their own criteria for what a point represents. I will not get into the specifics of a point (this is a topic in and of itself), but just know that points give the team a way to determine how much work is acceptable for a given ticket, and that any unit of estimation is fine. When the product owner has presented a story and the team has agreed upon the effort required to complete that story, then a software developer can take the story ticket and determine the subtasks required to complete the request.

I can see you right now, squirming in your seat as you await my key indicator for knowing that the amount of effort allocated to a ticket will be guaranteed to provide the quality that constitutes a win for your team. Unfortunately, there is not a direct answer for how grand or miniscule the scope of your tickets should be. A lot of the decision on acceptable task size is managed by the team, but the idea behind the bite-sized win is that you minimize the size and effort of each story and task as much as reasonably possible. Each ticket should be large enough to provide something fruitful for the clients and small enough that the team can celebrate work being committed to the repository frequently.

I have worked with many Scrum teams and we always have some sort of pointing system that is based loosely on the Fibonacci sequence: It usually starts at 0.5 points, representing the least amount of effort, then 1, 2, 3, 5, 8, 13 to represent proportionately higher amounts of effort; it can go as high as 20 points, though accepting a 20-point ticket is usually not advisable and warrants sending the story back to the ticket backlog, otherwise the team will need to drill into the request further and find a way to shrink it. A team will often create rules of their own regarding story sizes; perhaps a story with five points is too big, so the team will only accept stories that are three points or less.

Including the product owner in team discussions on desired levels of effort per ticket can help the product owner to focus on the scope of requested work in the future, and provide the product owner with valuable insight into how requirements should be presented to the team. As an example, say a team associates five points with a week's worth of work. When your team accepts a story that will take a week to complete, one software developer will be working on that task for one week, which means they will only deliver one body of work in that week. Before accepting, the team may analyze the requested work further and find that the five-point story actually yields multiple two-point (or fewer) stories. In line with the bite-sized win approach, by breaking down the five-pointer into several smaller-point stories, you ensure that tangible deliverables are made to the client more frequently and have less resistance. In Chapter 5: Working with Key Players, we will explore how bite-sized tickets can facilitate a positive working relationship with your client and others invested in your software development process.

The pointing system is a great way to learn how large stories are. When you and your team initially create a slew of tickets, it may sound like everyone has already done a great job of breaking down their stories into manageable sizes; that is, until the team is directly asked *how much effort* they believe each ticket will require, which is the question that arises in ticket grooming sessions. It is easy to look at a ticket on the surface and simply assume the requested work is small enough to be delivered quickly, but when you analyze the request, decide how much work needs to be accomplished before the work is considered complete, and apply the points, you may be surprised at how quickly that two-point ticket becomes an eight-point ticket.

A ticket-creating scenario

Let's explore a hypothetical scenario: we are on a software development team that is maintaining a financial system and our client wants to introduce a way for the consumer to make payments with their credit card when they complete a transaction. After working with the product owner to identify the effort required, we may decide to create an administration section within the software's UI, which is responsible for managing the credit cards the site will accept. We need to allow the client to enter their merchant information, add and remove credit card types as necessary, and manage which credit card vendors are to be accepted. We also need to add a dropdown list with the configured credit card types on the transaction pages. By the end of the project, we will have a fully delivered working credit card payment workflow.

If we were going down the old-school way of developing software, we would create an epic ticket, attach the nauseatingly long requirements document and write out a quick summary of what is expected, with a note that the software developer should refer to the document if they have further questions. The epic would likely include a list of any known issues ("verify this concern...") and link to all stories that are required to complete the client's request and make everyone happy.

What kind of stories would we create for a request like this? Here are a few to get started:

- Credit card management page.
- Access the API for each credit card vendor.
- Add a credit card selection dropdown on the transaction completion screen.

Most of these stories would be written by the product owner and maybe one or two developers would sit in on the conversation, just to say that a few subject matter experts were involved in the requirements-gathering process. In this method, the client only sees the final product but is never able to see the stages of software development as they happen.

We are better than this. We would like our tickets to exhibit bite-sized characteristics, and so instead we plan on delivering smaller bits and pieces, one at a time, rather than working in a vacuum for months, only to hop out like a beautiful butterfly and deliver a complete solution. How about some bite-sized tickets?

- Credit card management page:
    - Under-construction message, but still allow functionality.
    - Widget with tabs that show the acceptable types of credit cards:
        - Visa
        - Mastercard
        - Discover
        - American Express
    - Save credentials for each credit card type:
        - Visa

- ▪ [...and so on]
  - ○ API Integration:
    - ▪ Visa
    - ▪ [...and so on]
  - ○ Transaction complete page:
    - ▪ Add new credit card types to selection:
      - ● Visa
      - ● [...and so on]
    - ▪ Call API for selected credit card type after sale:
      - ● Visa
      - ● [...and so on]

The benefit to this type of approach is that the client can see what we are doing and raise any concerns. Perhaps, for example, the client sees these options and tells the team that they only plan on accepting American Express initially. Luckily, by catching this early, the team can reorganize upcoming work to accommodate the change before they put any effort into the other vendors, preventing them from wasting time on credit card types that are not required. Adding other credit card types can be a feature down the road if necessary. Any effort saved during development is a large gain for the client and the team.

In both the traditional and the bite-sized scenarios described above, the decision to omit certain credit card types may not have been discussed in the initial client calls; neglecting to provide details like this up front is a typical cause of software change requests. After all, the conversations surrounding altering software typically include a lot of topics that come from in-the-moment ideas. A lot of the time, the client does not know everything they will need until the software has been delivered. The idea behind the bite-sized win mindset is that iteratively delivering software via bite-sized tickets will provide all parties involved with a constant stream of information about what is new and what is being worked on. These bite-sized tickets should create a cadence of frequent deliveries that provides a greater level of transparency, rather than the client relying on scheduled status updates during the sprint. Uncovering certain nuances about the client's implementation ahead of putting the actual work in will ultimately save time and build confidence in the delivered software.

In the bite-sized scenario, we gave the client a series of small wins by showing them these little pieces, one at a time. Sometimes these pieces will have functionality, sometimes they are just visual, and either type allows the client to work a little more cohesively with your software development team to get the type of delivery that they are looking for. Throughout this book, we have discussed how clients are often left in the dark, and we have been arguing that anything the software development team can do to remove the mask of uncertainty will be appreciated by the client, who is now able to contribute to the process while deliveries are happening. Nobody wants to see the project fail, but without the proper processes in place, the ability for the client or the business to identify problems before they become larger issues is hindered.

Recognizing epics early

Within the Scrum methodology, the epic represents a large body of work that will certainly require a large amount of effort. For some teams, an epic might be a new feature in the software that may be developed over many phases. Others may only choose to label a feature as an epic if delivering that feature necessitates an increment in the software's version number. I like to define an epic as a project that will require multiple developers over an extended period. An epic is not a body of work that can be accomplished in a single ticket, but rather it should require multiple tickets to complete the request.

What if I told you that you can write a story ticket as though it were a miniature epic? That is, a story can be broken down into multiple tasks and those tasks can have multiple subtasks, which negates the need for an "epic" label in order to be treated as such. I like to use a series of smaller tasks to help myself properly define the elements of the overall stories. Breaking a story into smaller tasks allows me to take a request and isolate specific tasks that are required to deliver the final product.

To identify subtasks, I like to pre-groom my tickets before they are presented to the team in a grooming session. This seems like extra work, but by having my own pre-grooming sessions I can help to drive the wider ticket-grooming conversation. A ticket grooming session generally consists of all members of your team, and the focus is on the quality and content of the tickets. In the grooming session, the team checks that tickets include enough information that they will not have to request further information or research on the subject later, and make sure all outstanding questions have been answered to ensure the tickets are ready for accurate point allocation. A pre-grooming session is not a formal meeting in the Agile (or Scrum) process and is a meeting that I typically conduct when the areas of concern are isolated to just a couple of software developers. A meeting for pre-grooming is not mandatory, as you can pre-groom tickets by yourself without a meeting, or you can ask your software developers to pre-groom tickets in preparation for the formal ticket grooming session later. If no tickets require pre-grooming, then you never have to initiate the process. When I pre-groom tickets, I update tickets with the information that I have available; by the time a ticket makes it to the formal, broader grooming session, the team can then review my notes and add more context to the tickets or rewrite them with their own different understanding. I do not seek to define the tickets for the team, but rather prepare them for conversation. Subtasks can be added to the ticket at this time, but it is typically a waste of time because the broader conversation with the team may negate some of my assumptions.

Creating subtasks can also help me to understand if I am clear on the requirements and the effort required for the implementation. When tickets are given subtasks during the grooming session, you are often influenced by other conversations in the room, and you lose sight of your own personal experiences. If

I, however, can come to the meeting and tell the team that "Story X" requires six different subtasks and some of the tasks can be handled by another developer simultaneously, we can then use the time of the grooming session to decide if we want to share tickets and deliver sooner, and to let me know whether my subtasks are clear; it is possible that someone may receive my subtask and have no clue what to do, so the team may decide to eliminate some of my suggestions in the spirit of clarity or better defining the scope.

In this way, I sometimes like to think of stories as mini epics and find ways to break each story down into miniature features that can be delivered independently. Do not think of your stories as subtasks, since a subtask is part of the larger story that must be completed in full before the work is delivered. Instead, think of the practice of treating stories like mini epics as finding ways to break high-point tickets down into smaller, bite-sized tickets. Once you can confidently break down work, regardless of its size, you will be much more effective in assisting your team with preparing upcoming work. When I was in college, I always preferred working on projects that were delivered over the months that the class lasted, as opposed to a large project at the end of the year, which I would stress over completing the evening before it was due. It is no secret that there will be growing pains when changing your team's process, so you will need to set mini goals to gauge that you are on the right track; taking less time to scope and create tickets is a great indicator that your team is on the right track.

As previously mentioned, a feature is traditionally treated as an epic to help the team to isolate areas of work. Of course, most modern ticket-tracking systems will allow you to specify a *feature* as part of an arbitrary tag on the ticket, but some teams prefer to attach the centralized documentation for a feature to an epic ticket. When adopting the bite-sized win mindset, your team will need to examine all stories that come through and determine how much effort is being asked for each. If a series of stories, when taken together, is requesting a major enhancement to an existing feature, it could be that the series of stories combined constitutes a single epic and should be treated as a major project. On the other hand, several tickets that happen to be focused on a similar area do not necessarily make up an epic. For this reason, the team needs to work together to identify workloads and find a way to self-organize around the requested work. Using the bite-sized win mindset, it will be important to identify epics early and help the team to organize these tickets in order to properly plan how related work can be delivered.

Feature-gating

Feature-gating is a practice that allows you to control what your client sees. Sometimes, your software may have features that should not be visible to all clients, or they should not be visible until it has reached a certain stage of development. Using a feature gate gives you the ability to toggle the usage of code and/or the appearance of components on the user interface (UI) to hide it from or show it to the client. Feature-gating is nothing new, but it is nonetheless interesting, and there are a few ways to implement the concept of feature gates.

Many software development teams do not employ feature-gating techniques as a standard practice, due to either lack of knowledge around the concept or because their workflow deems it unnecessary, but it may behoove more teams to explore it. When utilizing bite-sized deliveries, because you are delivering small pieces at a time, you will sometimes introduce code that does not make sense in the whole and so should not yet be displayed to a user. Using a feature gate, anyone with sufficient access can disable its display until the feature is needed for review or has been completed and deployed.

Feature flags

One way to implement feature gates is to use *feature flags*: settings that tell pieces of code whether to toggle areas visible or invisible. Let's assume that a software developer has completed a bite-sized task that introduces a new, incomplete element on the UI. Perhaps the offending element contains a block of code that runs a sub-process that is still under review. No matter how you slice it, the new element does not work yet, so it does not make sense to allow a user to have access to it. This is where our new friend, the feature gate, will shine. We add some sort of flag to the database or directly in code, and then we use it to disable the element until it is ready for production. Sometimes, for example, a feature gate is just an *if(false)* statement that prevents execution in the production environment.

Feature-gating has many use-cases; it can be used to limit a feature's reach to clients, both before and after development, as we have seen, but it can also be used to customize software to each client's needs and avoid overloading them with features they will never use. I used to work for a SAAS (software as a service) company that created software for various short-term lending companies, and we would often receive feature requests for certain clients that were so nuanced that we would have no need to show them to other clients. For example, a very common request was the ability to create SQL reports that had business-specific requirements for the base criteria. In other instances, we would receive a request to integrate with a third-party vendor that nobody else would necessarily have a use for. We employed feature-gating for these purposes: we implemented the code into the software and applied a feature flag that could be toggled from a backend administration section; we could turn the features on and off as desired to control which clients were granted access. We found plenty of use in this practice because all our clients used the same version of software, but each client had their own database. We did not want to maintain a separate version of the software for each client, so by toggling certain features on and off, each client had what felt like a bespoke version of the software, without us having to create those different versions. These days, there are more efficient ways to manage features within the system, but our homebrewed version was fine for our purposes.

As an added benefit to the business, we could use feature-gating to charge clients a monthly fee for access to certain software features. This expanded our business to allow paid add-ons and the client could fully tailor their experience. For smaller clients, for example, it did not make sense to include a lot of the advanced features that our more-seasoned clients required. One of the biggest benefits of this process became part of my inspiration for the bite-sized win mindset, which I began teaching my software developers: while we were developing a feature for a client, we were able to give it to them in very small, bite-sized pieces and show them what was happening as development progressed, which allowed them to accept or reject the work before we had moved too far into the software development process.

On the flipside, we did find that the feature-gating concept turned into an over-utilized add-on fest, but that was only because we did not fully understand the concept. From a software development viewpoint, we disagreed with the over-use of add-ons because it required constant maintenance from us. We had to store data related to the features, ensure that we did not accidentally expose features during upgrades, manually test with the features enabled and disabled, debug when a feature was enabled and later disabled, and a slew of other headaches that came our way as we were asked to add new features that would be hidden behind feature toggles. In this case, we had to accept the overuse of feature gates, because it turned out to be a major financial benefit to the business.

A/B testing

Another variation of feature-gating is the A/B test, in which we directly compare different versions of something like a web page or software layout elements, and control who sees what version. For example, I used to work for a marketing company that focused on website marketing. We were developing different versions of our website and needed to know which versions were effective, so we needed a metric to use to compare versions. In the world of web marketing, you have two main concerns: clicks and bounces. A *click* is when a user navigates within your website after they saw your initial landing page; you consider the click to be a small win, especially when those clicks convert into a sale. A *bounce* is when someone visits your website and, within a certain amount of time, leaves without doing any actions, such as clicking or filling out a form. A bounce is typically measured as a few seconds or less, but it really depends on the company and how they wish to view it.

At this website marketing company, we used A/B testing by segmenting the traffic to our website and redirecting different users to different versions of our homepage (or whatever page we were testing). We would direct a certain percentage of traffic to each page. *Percentage-A* would see *Version-A* (likely the original version) of the homepage and *Percentage-B* would redirect to *Version-B*. From there we were able to track how many clicks we got and view our conversion rates to know how successful our changes were. We could also answer other questions, such as how many users were bouncing immediately and how long our visitors would stay. Over time, we could identify which elements of our website were the most effective at getting users to sign up for our service. Sometimes, we would want users to access our self-help sections, so we would try to tailor our homepage to that type of traffic. This testing method was great for when we ran ad campaigns and wanted to see which landing pages had the best conversion rates. Web marketing is intense, so it takes innovation and a willingness to experiment in order to succeed.

If we think about A/B testing as feature-gating, we can describe the process as showing different features of the website based on the desired website traffic. In the example, the entire homepage is one feature that we gated by driving or restricting traffic to it. We were able to toggle different versions of the website based on who we wanted to see it. Our testing was traffic-based, so backend logic decided who received the associated features. Importantly, if we already had an established userbase, we would be more careful about which users were affected by the desired changes so as not to interrupt their expected experience. Using landing pages for A/B testing will often yield decent results, because an ad campaign sends a target audience to a desired page. If you ran an ad campaign that was targeting middle-aged women, you would handle your A/B testing differently than if your target audience was college-aged men.

Summary

As previously discussed, when your team achieves a bite-sized win, your client and the business achieve one too. When implementing bite-sized tickets, the team will benefit from a reduced scope during their software development cycle, and they will see progress made. The client will benefit from being able to reduce or increase the scope of their request before development has progressed too far. Epics are a great way to organize many stories into a single group, but do not be afraid of rewriting your tickets to be smaller if it guarantees that more developers can work on the same story. It goes without saying that you should always give your clients regular updates about progress of the work that is being performed; these bite-sized wins can be served as updated software, which the client can play with as opposed to verbal updates where your product owner attempts to paint a picture about what has been completed. Working with your team to implement the bite-sized win mindset will require a great deal of focus and organization, which will partly rely on how well your team is able to author and organize their tasks.

Feature-gating is a useful tool for software teams. The product owner will often need to hold a demo session for the client when work is complete, so it helps to be able to give them a way to toggle the work that is in-flight. When the feature has reached a stage of maturity where it is displaying and manipulating data, the client will find value in being able to see the feature displayed against their own data. If implemented correctly, we avoid giving clients software cluttered with half-baked features and software developers can continue to develop without worrying about accidentally displaying an unflattering component. There are plenty of tools out there for implementing feature-gating, and it is not a difficult thing to implement for yourselves.

UNCOVERING CERTAIN NUANCES ABOUT THE CLIENT'S IMPLEMENTATION AHEAD OF PUTTING THE ACTUAL WORK IN **WILL ULTIMATELY SAVE TIME AND BUILD CONFIDENCE IN THE DELIVERED SOFTWARE**

Chapter 4 | The Toss-5 Method

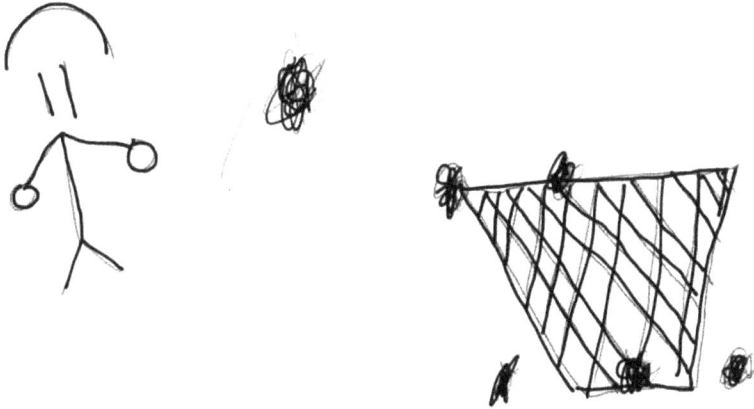

This chapter will discuss another, perhaps more fun way of implementing the bite-sized win: the Toss 5 Method. This method will put you into the mindset of recognizing and seeking bite-sized wins through daily repetition. As we have discussed, it is important that you receive miniature wins daily and use them as fuel to motivate you. While I do not stake claim to the method itself, it is an excellent example of the bite-sized win mindset.

What is the Toss 5 Method?

I have a reminder on my phone that pops a message on my screen every morning at 9:00am for a whole month, which contains only two words: *Toss 5*.

When I see this message, I look around my desk and pick up five things that I no longer need in my life. If the number of items in my hands is less than five, I scour the office and then the house. These items could be anything, as small as a wrapper that I forgot to toss or as large as an old knick-knack that I broke ages ago and just never threw out. When I have finally picked up five insignificant items, I open the garbage can, toss them in there and never speak of them again. I do this every day for a month. The feeling that I get after I declutter my life is incredible. We gather items in our lives every day and rarely stop to think about the consequences of holding on to them. Some people let this clutter build up for years. It is important to rid your life of the things that you no longer need,

because you are inevitably going to gather more items, and you never want more clutter in your life than you can deal with at any given time.

This process is known as the *Toss 5 Method.* I was not the inventor of the Toss 5 Method, but I have been a great beneficiary (from what I can recall, chatter around the Toss 5 Method started quite a few years ago when minimalism was at its peak). Whenever I feel like my life has become cluttered, I go through this process. Think about it: if you toss five items every day for a month, by the end of the month you have ~150 fewer items cluttering your life. Your home is tidier, and you feel a lot happier. If you are someone who has a hard time letting go, turning the act into a daily process is a great way to get around that feeling. Completely upending your life and discarding everything that you own in one sitting can be downright stressful. If you need to only ask yourself to find five items every day, you are more likely to achieve your goal of a smaller footprint.

So how can we apply this concept to software development and (more importantly) the bite-sized win mindset? Consider the following chapter as a challenge for you and your team in your professional endeavors. Tiny goals with real results, like tossing unneeded items, are at the core of the bite-sized win mindset and if you can convince your team to join you, they may start to crave those little endorphin boosts that come with each win. Let's explore some ways that the Toss 5 Method can be relevant to software development teams.

What are the rules?

In a general sense, there are not many rules to this challenge: we just pick five things and toss them, right? But let's broaden the definition and add a little structure for the sake of the challenge. Our goal is to declutter our lives and improve ourselves professionally. We want a way to receive small, daily wins that encourage us to build the habit of seeking more bite-sized wins on a regular basis, and we want to encourage the team to think about goals as small and achievable, as opposed to large and infrequent.

Without further ado...

1. **Choose an activity to do each day.** Are you de-cluttering your working environment? Then toss insignificant items. Are you cleaning up random bits of inefficient code? Then refactor lines of code. Do you want to add subtasks to tickets in the queue? Then add subtasks every day.

2. **Decide how many times to do this action.** *Five* is merely a recommendation. You may want to delete ten emails each day, or maybe thank two coworkers for something they have helped you with. The sky is the limit, and your options are only limited to tasks that you feel are small enough to tackle at one time and that still encourage you to maintain this habit long enough to overcome obstacles.

3. **Set a reminder on your phone.** Think about how many days you want to do this challenge for. If you are de-cluttering your office,

it is not realistic to continue daily for the next two years. Nobody has that much junk in their office (*I hope...*). You can also decide if this is strictly a weekday activity or if you can do this over the weekend as well. Whether you do it in the mornings or at night may not make a difference, but try to avoid doing this during lunch, since that time should be spent recharging.

4. **If you meet your goal early, end the challenge**. Let's say you want to add three subtasks per day for the next twenty working days, but you find that there are no more tickets remaining after the sixteenth day. Then, by all means, you may end your challenge early. There is no need to push beyond what is advantageous.

5. **Hold yourself accountable**. Maybe get a coworker or even the whole team involved in this goal, then check in on each other every morning to make sure you are all on track. Since this is about bettering yourself, why not have your team benefit as well? Just make sure that whatever goal you choose, you and your support system are prepared to see it through until the end. Do not pick anything that needs more immediate attention than the system can provide or that you know you will not stay honest about completing. We want bite-sized wins, not bite-sized failures.

Refactoring code

Keeping with the theme of relevance to software development teams, I suggest refactoring code as one goal for this challenge. Say you have a coding standard that you want to implement across the project. Scanning the entire codebase for one poorly formatted line is time-consuming, but when you break your code repository down into bite-sized chunks, you can check one chunk at a time, and you may find yourself more willing to put forth the effort. For example, if your coding standard is broken due to too many levels of nested code, you may not be able to find all occurrences of over-nested code in one go. Instead, challenge yourself to scan five files every day until you have searched every file in your repository. On the inverse, if you have a pattern that is easy to find, execute a "find all" command across your project, jot down every occurrence of the pattern and fix five files each day.

You might also write those outstanding unit tests you have yet to finish. Each day, write five important, or even inconsequential, unit tests and, over time, you will increase your code confidence. We are not all fans of test-driven development (TDD), but if you are writing tests after-the-fact, you may well get a better understanding of how the current code works. Be aware of time requirements: a unit test may require some code refactoring in order to get it in a testable state, so try not to over-commit yourself. You could find that requiring more than two or three tests to achieve the daily win ends up discouraging the team.

Tickets

When your team is knee-deep in a high-priority feature, you understand the value of having tasks prepared to keep the team on track. That is why it may be a good idea to focus on tickets for your Toss 5 challenge. Subtasks are a great way to help the team understand how many points each story should be allocated, and to help your fellow developers to tackle

their tasks in a calculated manner. Your challenge could be, for example, to identify the set of tickets likely to be assigned to a particular software developer who is the "expert" on that subject area in your codebase. Break out five subtasks from that set and give them to a junior software developer to guide them on how to complete the task. These subtasks will likely require the junior developer to write code, so make sure you include as much detail as possible. Put yourself in the junior developer's shoes and ask yourself what would have helped you to be successful early on in your career.

Another great way to focus on tickets during this challenge is to commit to finding loose and previously overlooked requirements and working with your product owner on fixing them. You might commit to investigating one ticket each working day for the next month, ending up with twenty tickets that now have solid requirements that your team is more enthusiastic about accepting into an upcoming sprint. You do not even need to commit to resolving the tickets. At a minimum, commit to *reading* one upcoming ticket per day just to be aware of any nuances that the team should consider.

Your working environment

Everything comes full circle, right? Applying the Toss 5 Method to your workplace can help clear your thinking space. My home office was pretty much the reason why I started doing this challenge a few years ago. I get to the point where the office is cluttered, and I lose control of it. I spend most of my time with my head down in code, so I do not notice when my mail pile has grown to be knee-high. When I apply the Toss 5 Method, I like to commit to tossing five items every morning for a month. I have managed to gradually de-clutter my house on several occasions using this method.

It is hard to decide what to toss when you feel overwhelmed and want to toss everything (or nothing). It can be difficult to let go, but I often find that by the end of the first week, I am excited to throw things away because I am seeing the benefit of less clutter in my home. I sometimes even break the rules and toss more than five things in a day, just because I have the energy to do so. You oversee this challenge, so break your own rules occasionally.

Involving the team

This exercise can easily become fodder for promoting team engagement in the process. Work directly with your team in order to find out what needs the most attention, and what might benefit from the Toss 5 Method. The goal is to find some bite-sized wins that will keep the client happy and reassure the business that you are a capable and competent team. Find out if there are legacy features that have been sorely neglected. Find out what is coming down the pipeline that requires particular attention.

Some software developers do not enjoy formal teambuilding exercises, so forcing involvement is not the way to go. The Toss 5 Method can easily be presented as a suggested daily task. The end goal should be clear and the expectation for the team should not be a mystery. Refactoring and ticket grooming, for example, can easily become a routine part of the software development process.

Summary

The Toss 5 Method is not a new concept, nor is it exclusive to software development. In my opinion, the greatest benefit to trying it out is that it gets you used to bite-sized wins. As software developers, we should find a way to use this method to our advantage. Whether you are refactoring code, finding missing unit tests, authoring better tickets, or just cleaning up your office, the Toss 5 Method may have a place in your arsenal when teaching others how to implement (and appreciate) the bite-sized win mindset.

TINY GOALS
WITH REAL
RESULTS
LIKE TOSSING
UNNEEDED ITEMS
ARE AT THE CORE
OF THE BITE-
SIZED WIN
MINDSET

Chapter 5 | Working with Key Players

In the pursuit of implementing a bite-sized win mindset within your organization, there is a high likelihood that you and your team will run into some obstacles surrounding acceptance of the new workflow by others who share an interest in your project. It is important that other people within and outside your organization understand that the benefit of accepting the bite-sized win mindset into your daily process is constant feedback to all involved, including the client and the business.

The goal of this chapter is to discuss the different groups of people involved in your software development process, how they might oppose your bite-sized win mindset efforts, and how you can win them over. We will explore the different *key players* within the organization, who they are and how they might create roadblocks for your team during their software development journey. Specifically, we will discuss the client, the business, and your software development team, and how to handle objections and misunderstandings from each. Prior to this chapter, we have explored what the bite-sized win mindset is, but now it is time to find a way to bring the workflow into your team's software development process.

The bite-sized win mindset is primarily about improving the psychological aspects of the software development process: it strives to make everyone involved in the process happy by highlighting progress and creating an avenue for intervention before the final delivery. You should also start to see some improvements in your delivery times, as well as your team's ability to identify issues early. While your focus may be on improving transparency, you are going to notice that other key players are more concerned with the amount of time that the delivery takes. Keep the concerns of all members in mind when you make any change to processes. When initially selling the team, the business and the client on adopting a bite-sized win mindset, you will want to place emphasis on the benefits of the process, take baby steps so as not to cause disruption, and make sure everyone is focused on the same goal, even if it means combining your own goals with their goals. The most important thing to remember is that bite-sized wins are more focused on improving transparency than on improving delivery times; however, improved delivery times will often be a positive side effect.

Any of the key players may prove to be a roadblock in adopting a bite-sized win mindset. Clients are focused on deadlines. The business is worried about making a good impression on the client and convincing them to provide more work in the future. Software developers just want to do their job the same way that they have always done it (*if it is not broken, do not fix it!*). Let's explore how these key players fit into the new bite-sized win workflow and how we can work with everyone to ensure a smooth, successful transition.

Clients

The level of involvement your client has in your software development process will dictate the level of effort required to get them to buy into adopting the bite-sized win methodology. Some clients like to remain hands-off and are only concerned with delivery dates and the amount of bang that they get for their buck. You would handle obtaining their buy-in differently than a client who is very hands-on. You cannot control the level of involvement coming from your client, nor should you try, but you can explore approaches to dealing with both possible client types.

Hands-off clients

Hands-off clients typically meet with your product owner on a predefined schedule to ask for status updates and ensure that the planned deadline is still realistic. They are not very interested in how the work is being done, so attempting to have a technical conversation with the hands-off client will likely go nowhere. The hands-off client focuses their energy on submitting requests, giving deadlines, checking up on progress, preparing for potential delays that may occur and ensuring the project stays within budget. The amount of time that it takes for their project to be delivered will dictate how receptive they are to the process changes your team makes. They may be hands-off for any reason, such as inexperience with software development or because they are simply too busy to intervene. Respect their distance and make your team's decisions based on this knowledge.

When dealing with clients who are not hands-on in the software development process, I typically do not inform them about major changes in the process, but simply ensure that I keep them up to date with the progress of the requested work. When the workflow for the project becomes geared towards the bite-sized win mindset, your hands-off client will often find themselves more interested in seeing updates as they occur. In a traditional workflow, the client will structure their expectations around the results of status update meetings, in which they receive new information about roadblocks related to their request and work with the team to overcome them. In a bite-sized win workflow, they may instead become interested in the implementation as it is being developed and may start to think of potential issues and discussion points before the meetings, giving you more time to develop solutions and discuss requirements. The change in process should help your hands-off client to become a bit more hands-on and can help save time in meetings that you would usually dedicate to updating statuses.

If you are starting a project and are very early in the process of adopting the bite-sized approach, you might find that you are a little slower to deliver, but you will pick up momentum as your team becomes comfortable with splitting up tickets and working with smaller requirements. However, since the team has not established a delivery cadence with the client, the hands-off client will not notice and will not likely care. What the client will see is that, from the get-go, the software development team delivers work frequently. The hands-off client does not notice the effort involved in switching to bite-sized deliveries, since they often lack a frame of reference.

If the project has been underway for a while and the deadline is still in the distance, a change in delivery cadence caused by the bite-sized approach will be recognized, even by the hands-off client. The client may be used to biweekly or even monthly deliveries. From the infrequency of deliveries, the client has established that your team requires minimal attention, so they may not adapt easily to more frequent deliveries. This is not to say that they will be *upset* about more frequent deliveries, rather I have noticed that the hands-off client will not always take the time to review the more frequent work as it becomes available. In this scenario, your team will need to acknowledge that the increase in delivery frequency is more of a benefit to the team and to the business than it is to the client. When the hands-off client is ready to analyze the work, it will be available to them. Until then, the work will sit dormant. Later in the process, the hands-off client may start to pay attention to the work as it becomes ready; this typically happens closer to the deadline, when they want to take a more active role in what is being produced.

If you are working on a project that is coming down to the wire on time, it may not be a good time to change the process or delivery frequency. Save it for another project. Disruptions in projects nearing completion are likely to upset your client. I know that I am trying to sell you on this idea, but in the end, your client is your bread and butter. If you upset your client, they may well leave, and without clients, you will not be able to fund your organization. Use your best judgment about whether to change your process or just finish out the project and worry about a new software delivery process in the future.

Hands-on clients

The hands-on client is involved with the software development process itself. Perhaps they already have an in-house software development team, and the agreement is that your team would work with their team. Perhaps they just have an overly curious mind and enjoy receiving frequent status updates. These clients are generally a bit easier to work with. However, the buy-in for adopting a bite-sized win approach from the hands-on client may be more difficult to obtain initially, since they feel they have more control of the project and may not generally want to share that responsibility. Having said that, if you can explain to them in layman's terms that your goal is to divide work between software developers more evenly and create a cadence of more-frequent software deliveries, they will generally be receptive of the proposed changes, especially because it will give them access to the current state of the project without having to wait for conversations from the team every time they are curious. The team will need to commit to transparency during the changing process and answer any questions that the hands-on client may have.

In my experience working with a hands-on client, if the project appears to be at risk of missing a deadline, the client will work with the team to rectify the obstacle or extend the deadline to a more realistic date. The hands-on client will recognize when the delay is not caused by the process change. Their inclusion in the process often generates a sense of pride and excitement, and they feel partly responsible for the success of the project, whether it was completed on time or after a slight delay. This is not to say that the hands-on client will always be accepting of a delay, but they tend to be a bit more empathetic than the hands-off client, whose distance from the process means they often lack insight. The hands-on client will also pay attention to the warning signs of an impending delay and, if their project allots the leeway for a shifted due date, is more receptive to pushing the deadline out. This especially holds true if the hands-on client has been observing more frequent deliverables, even if the deliverables are iterative and not completed. The team and the hands-on client can work together to formulate a plan around what changes are necessary to accommodate the new timeline.

One major difference between the hands-on and hands-off client can be seen when the deadline is coming up for the project. If it seems as though the deadline will be missed, the hands-on client will often want to come up with the solution on their own, whereas the hands-off client will often expect your team to decide how to operate to ensure deadlines are not missed. If the deadline is in jeopardy, the hands-on client may still be receptive to changes in the workflow, assuming you start the conversation about process changes as soon as the deadline issue is identified. A project in jeopardy can invoke a panic mode in which the client is willing to do absolutely anything to ensure that the project is delivered on time. In this circumstance, switching to more of an iterative, bite-sized win mindset can be of benefit because you and your client can easily highlight what work has been completed and then the team can decide on the best way to pivot the planned approach in order to complete the project on time. If the client has approved of what has been delivered in past sprints, they may opt into a *phased launch* process, in which you release your project in versions, such as "Version 1.0, Version 1.1" and so on. That way, when the deadline rolls around, you have something to present, and then you can build on the earlier versions to complete the full work after the deadline. This suggested change in the software delivery process has its pros and cons, but there is no one-size-fits-all answer for how you and your client should approach a jeopardized delivery date. You should know your client and understand any risks to the project.

Early in my career, I was part of a very small software development team tasked with changing the main website for our company. The company's owner was of the hands-off client variety. While he did sign our paychecks, he did not inject himself into our daily work. The buzzword "Web 2.0" was flying around, and the company's owner was anxious to have this technology implemented, but none of us understood what it meant. We decided to incorporate new design ideas and modernize the old website, because that was how we interpreted *Web 2.0*. For example, we went from using sharp-cornered boxes to using more aesthetic rounded buttons. This does not sound very innovative by today's standards, but back then, buttons were all rectangles and if you wanted round corners, you had to create the graphics. Without bogging you down with too many details, I do mean it when I say that these websites required a *long* time to develop, and our six-month timeline was tight. I am a huge proponent of phased implementations, and this is the approach that we took when we realized that we could not complete all the requested work by our original deadline. For much of the website, we did not implement full features, but made sure to introduce enough of a feature to allow the visitors to achieve what they came to do, giving us more time to complete the full work. This phased approach gave us the added benefit of tracking trends with our website visitors to determine which features to spend more time developing. There is a common phrase "data is king", and this was a concept my company felt strongly about. Our phased approach gave us the opportunity to use the data to figure out if we should invest more time in our website or hang our heads in shame and go another route. Spoiler: We were successful, and everyone loved us.

Working with the client

Whether your client is hands-on or hands-off, your team will need to gauge their level of commitment when updating any software development processes. When your team decides to move to the bite-sized win mindset for creating tickets, for example, you should not inadvertently introduce a series of bite-sized losses that might deter the client and reduce confidence in your team's ability to deliver high-quality software in a timely manner. As my experience has shown, working with smaller deliveries early on will potentially have a positive impact on the full software's delivery, due to the increased transparency. Throughout my career, I have heard lots of people with fancy job titles mimic the phrase, "plan for the unknown." While that seems like an impossible feat (*how do you plan for what you cannot plan for?*), it is important to consider that your actions can influence your ability to deal with issues; will some left-field obstacle be easy to overcome, or will the obstacle turn into a brick wall that cannot be passed without immense effort?

Here are some common problems statements that may come from the clients surrounding process changes, and some ways to handle them.

Problem Statement	Solution Response
I do not like change.	This project could use a fresh coat of paint for the software development process. This change will expose areas within our process that require additional attention.
We have been doing fine so far.	Our teams have been unable to identify potential problems with requirements until the work has been completed and deployed to a staging environment for your team to review. We would like to minimize this risk and make sure that we always deliver the desired outcome for your team and clear up any confusion ahead of time.
Will this speed up delivery?	Not at first, but once the team is comfortable with the process, we expect to see a decrease in the amount of time that it takes to address requirements that require tweaking.
Can we go back to the old way of working if this does not work out for us?	We can always go back to the previous workflow. Any work that our team has completed should remain as-is instead of reverting. However, this process takes time to implement and more time to see its effects. Expecting an end date to the process is not the best mindset to go into this with.

Business

I have frequently made references to "the business" throughout these chapters, but I have not yet taken the time to explain who that refers to. In the spirit of a bite-sized definition, "the business" shall henceforth be defined as *any person or entity within your organization who is not directly involved in the software development process.* This might include any of your executive-level staff, anyone in your company's client service or support teams, the marketing department, or any other department who does not work on gathering and delivering requirements to the client. Depending on how the organization is structured, the business could include your product owner and product team (which happens often in smaller organizations), though if that were the case it would likely benefit your organization to work as closely with the software development team as they do with the product team. If you include more people from the business than just software developers and product owners in your team, then you have more eyes on the process and greater support as the team grows and matures. The business plays a pivotal role in your team's success and ability to operate.

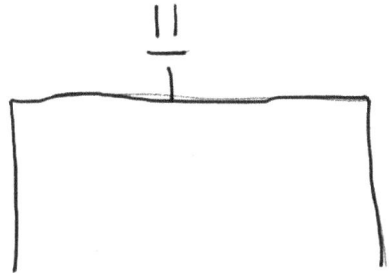

When coordinating with the business, it is important to remember that their top priority is the client and ensuring that the client is happy enough to remain a client in the future. When approaching process-related conversations with the business, then, you need to remind them that the change to the bite-sized approach will benefit the client just as much as the overarching organization; a change in process is not only for the benefit of the software development team, since at the end of the day, the team is able to carry on with any old processes. The driving purpose behind updating a team's software delivery process is to build the framework for managing future projects.

The bite-sized win mindset aims to close gaps in client communication and eliminate the possibility of inadvertently leaving the client in the dark about the status of the work. By using the bite-sized approach to allow anybody (with proper access and resources) to see deliverables occurring at a greater frequency, the key players will start to have more confidence in the process itself and see the advantages in its efficiency. For example, say your team has a process which allows the client to check up on progress on a weekly basis. During this check-up, the client will likely take every liberty available to inquire about updates, ask for demonstrations, question roadblocks, ask for changes, clarify unclear requirements, and drill into every possible issue that may have been sitting on the table since the last meeting. Those are a lot of potential conversations. If the client only has a 30-minute meeting once a week, your team will not be able to cover all topics necessary, leaving your client dissatisfied or questions unanswered. Therefore, it is more desirable to have updates constantly available to the client in such a way that they can answer their own questions and reduce or eliminate the need for scheduled meetings. You can implement a process that exposes the team's status in such a way that your client can find the information they are seeking without your intervention. Once the client is confident that they know the status of the work before it has been officially presented to them, your conversations will be more fruitful and have clearer direction.

However, keep in mind that by increasing the cadence of deliverables, regardless of the size, you can end up spoiling your client, and they will expect these deliverables to continue happening at this frequency. *Imagine their surprise when the wins never end.*

When planning for adopting the bite-sized win mindset, you should carefully consider how you will deal with the new delivery cadence. The business may be used to a certain delivery schedule and may have specific processes built around it. Make sure, too, to be consistent in cadence changes. It will not impress anyone if your team adopts a process of weekly deliverables and then randomly misses three weeks. I would suggest that if your team currently delivers software once a month, work with the business to test delivering once every three weeks, then two weeks, then weekly and finally commit to a cadence of delivery twice a week, or whatever works for you and the business. Of course, I am not suggesting moving to continuous delivery, as that would often be the responsibility of another team within your organization, but I am suggesting that you stop your team from being a roadblock from letting that happen. If your team were to set an overly frequent delivery schedule that they could not practically commit to, the result could be a loss of confidence from the business, which the team had worked so hard to build. Creating bite-sized stories should be implemented in such a way that the business can be confident that the client is seeing progress and the process is not jeopardizing the existing system that is in place.

When approaching a new project, consider your process. Your team will start with a blank slate without requirements on how frequently work should be delivered, since the software will not yet be in use. This is an ideal scenario for implementing a bite-sized win mindset: the business likely has no expectations on how frequently the client will receive deliveries, so your team can create a pace of consistent deliveries that encourages the client to provide frequent feedback to the business as they view the work that has been delivered. If it turns out that your team is a bit too ambitious with their delivery frequency and needs to scale back, the business will likely assume that the initial flood of received work was just the team getting set up. Over time, the team will find a rhythm and the business and client will continue to enjoy the bite-sized wins. I always caution teams to be delicate while making changes to their software development process. If your new process fails, it will be noticed very quickly by the business, but if you introduce changes to your process slowly enough that the business barely even notices it was happening, you may experience less resistance. Be mindful that if your team is attempting to make several large changes to the software development process at one time, you will likely experience a great deal of resistance.

Another way to successfully pitch this proposal to the business is to focus on the benefits of improved communication with the client. Now that the client has enough constant information to answer their own questions, the business may find that scheduled conversations can get to a deeper level, beyond the typical status updates. It may be that currently the client and business need to go through several software developers finding out who has the latest update; this holds especially true for tickets that have been broken into subtasks that were subsequently handled by more than one software developer. With bite-sized deliveries, status updates are easier to see since the work in question will likely already be checked in and available for demonstration. Even if the request is not ready for final delivery, smaller tickets will be checked in faster than one main ticket with several subtasks. The client will see the changes as per your delivery schedule, but the business just needs to log in to your staging or development environment to see deliveries happening in real-time. It is easier for the business to see updates and then update the client on the deliverables that are ready to go than for the business to always wait for updates from the development team, especially when tasks take several weeks (or even months) to complete.

To keep the business happy, make sure that your team has identified whether any process changes will affect the business and has done enough research to ensure that the changes will not have a negative impact on the business. The business is primarily focused on the image of the organization, and your ability to reliably deliver software does not just reflect upon your team; nobody outside of your organization views your team as being a separate entity from the organization. With enough planning, your team can find ways to start slowly implementing bite-sized stories into your workflow.

In one company I worked for, the business and the software development team did not see eye-to-eye on modernizing our software development process. We wanted to implement smaller deliverables, but they felt that the current workflow was sufficient. The business did not have enough confidence in us to let us make changes without their oversight, and they created obstacles around every turn. We had to reassure the business side that *we* were confident in the process changes while also convincing them that our changes would be beneficial. One way that we did this was by sending our team's leaders to obtain certifications in the Scrum methodology. Having a certification helped the business to see that the ideas were formalized and accepted within the industry. We also gained confidence by slowly incorporating what we learned into our process. We kept our changes bite-sized and ensured that big changes were not instant and would give everyone involved the ability to request changes before anything was agreed upon. Without question, it will always be easier to work with the business when your team has a positive relationship with them. However, if the business views your team in a negative light, there is still a road to redemption: the business will simply require a greater deal of handholding throughout the changes, and your team will need to actively work on convincing them that the updated process is a net positive for everyone involved. Do not lose sight of your goal and ensure that the path to delivery is still clear.

The bottom line is that you and your team should exercise caution when implementing new processes, whether they are for your software development process or otherwise. Make sure that you always keep the business in the loop about any obstacles that you may run into. You should acknowledge and respect that the business and the software development team are on the same side, and you all want to deliver high-quality software that everybody is happy with.

Here are some common problems statements and solutions regarding implementing the bite-sized win.

Problem Statement	Solution Response
I do not like change.	We are currently unable to provide adequate updates to our clients. Modernizing our process will allow us to have more fruitful updates for our clients when they ask for a status update.
We have been doing fine so far.	The current software development culture is moving to more of an information-on-demand style, and our clients will appreciate that we are not using antiquated processes to develop their software.
Will this speed up delivery?	Our top priority is to ensure that our clients are not left in the dark during the software development process, and this updated process will allow us not only to provide them with more accurate updates, but also to provide them with hands-on testing of their software as development progresses.
Can we go back to the old way if this does not work out?	When our clients begin to see results more often, they may become hesitant to move away from this modernized process for developing software. It is important that they can have access to frequent status updates, and they will find great value in being able to chime in on potential issues as we iteratively develop new features.

Software Developers

As a software developer, I can confirm that software developers do not like change. In fact, we have a strong preference for routine. Typically, when you approach a software developer and reveal that you have a grand plan to upend and fundamentally change the software delivery process, you may initially be fooled into thinking that you have received immediate buy-in from them. They may seem accepting, willing, and ready to go. I am sure that after your spiel about how well the software development process will be improved, however, they smiled politely and immediately thought, "Good luck." Once your attempted process change is underway, reality will set in, and getting your whole team to go along with it will likely feel like pulling teeth.

Obtaining initial buy-in from your software developers might be the easiest of all the key players, but you will quickly learn that they require the most supervision. Your software developers may be nonconfrontational, but they are the most likely to quietly slip back into their old habits, quite possibly without you noticing. It can be difficult for software developers to break their cycle, as I have seen firsthand when updating any piece of the software development process. For example, when my team was moving our ticket system from a home-brewed solution to a commercial solution, some of our software developers continued creating and maintaining tickets in the old system and simply left notes in the new ticket system that referred to a ticket number in the old system. They were not trying to be defiant; they were just used to working with the old system and knew the steps necessary to create and manage their tickets. Software developers enjoy their autonomy, so it may be difficult to break their habits.

Perhaps you are familiar with the old trope, "It takes *x days* to break an old habit." I do not subscribe to this belief. Habits are not necessarily broken by a set number of days, but rather by the *repetition* of new habits and the commitment to changing habits. If you find your developers slowly slipping back into their old ways of working, you will need to give a friendly reminder about the updated processes. Software developers tend to be very robotic, and you may find yourself reminding them a few times before they get it. Do not shy away from giving friendly reminders, but be careful not to overdo it and risk alienating single developers.

You should also involve your team in the process change. When a software developer is involved in changes to workflow, they often have an easier time investing in it and following it than software developers who are told to blindly follow a new software development and delivery process. Make sure to take the time to explain why a process change happened. It is natural for people to assume that they know best, so without proper explanation, they may not understand the benefits.

One benefit you might expound upon is the opportunities for cross-learning. At one financial company I worked at, my team decided to change our process to focus on delivering smaller units of work (just like the bite-sized win mindset). The driving factor behind this decision was that we discovered that our software developers had been working in "silos" and were not collaborating on the work that was delivered. The result of having siloed software developers is that nobody was being cross-trained, so everyone was pigeonholed into a specialized section within the software and when it was time to take on new work, we had to evaluate the workload of particular developers who had the right skillset for those specific tasks to make sure they were always assigned to their specialties within the software. As you can imagine, this created a bit of a nightmare scenario in which our product owner could not give high-level delivery estimates to our clients unless he spoke with our development lead and manager beforehand, since a conflicting high-priority task that was already assigned to a "pigeonholed developer" could be slated to consume their time for the next few sprints. Our product owner did not have insight into who was pigeonholed into certain areas. When anybody wanted to learn something new, such as a new technology, they were limited by the requirements of the section of software they were stuck developing. While it was nice that we had software developers who were experts in certain areas, it made for a heavily restricted workflow when we wanted to deliver large amounts of work in a short amount of time.

Before the proposed workflow change, we worked with larger units (stories that included full functionality) and we would not check-in code to the main repository until all work on a unit had been completed. The common practice was for a software developer to take a task, disappear for a week and then hopefully resurface with a finished product. When we modernized our workflow to break these stories down into smaller tasks, we discovered that our software developers were still taking a few stories at a time to make up a larger story, even though the units of work were broken down into smaller chunks. We called this practice "hoarding tickets." They were not turning in their work until their full stories were completed. This meant there was really no difference in the old and new workflows, which taught us a very valuable lesson: when assigning work, we had to make it clear that one developer works on one ticket at a time. In order to enforce this, we updated our team's delivery agreement to state that tickets in a sprint would remain unassigned. A common practice in Scrum (or any Agile methodology) is to find a team's velocity. Velocity is calculated as the number of points that the entire team can complete per sprint. In our case, we were originally looking at velocity as being per software developer, as opposed to the entire team, which was the incorrect method. Once we were able identify a consistent range of points for the team's velocity, we were then able to plan for that number of points per sprint. When a ticket received points, we would not assign the ticket. Instead, every software developer worked from our team's sprint board and only took tickets that were unassigned, which differed from our old practice of pre-assigning tickets before the sprint began. We would rank the tickets based on priority, and when you finished working on the ticket and checked it in, you would take a ticket from the top of the sprint board. If the ticket on the board had a dependency on another ticket in that sprint that had not yet been completed, the software developer could skip it. Our software developers finally had a set direction and a clear and easy way to adhere to the new workflow.

The lesson from this experience illustrated a very important concept in software development (and information technology in general): *separation of concerns*, which is the notion that one idea produces one unit of work. Translated into the realm of coding, this usually refers to how different sections (or different modules) within the software are responsible for only one goal (a concern). For example, if a software developer accepts a task to update a table in the database, they should not also alter the backend code to reference that table's new update; the first concern is the database and the second concern is the backend code. The point is that you start to separate your work into segmented tasks that do not overlap with each other.

In terms of the bite-sized win mindset, we are assuming that each bite-sized unit of work is producing one concern. When our software developers would hoard multiple tickets, they viewed it as being concerned with one project or one feature. While this is true in some ways, they were thinking too large and missing the point: each ticket was its own concern, so each was structured to be its own shippable task. It took some time to teach our software developers this new way of thinking, but once they saw the light, I truly believed that the developers were happier. The moment of clarity came for them when they noticed that work was shipping faster, and changes made to the requirements took less effort to respond to. They no longer had to spend months working on a new feature that was destined to be altered several times throughout the process. Chapter 6: Separation of Concerns will dive much deeper into this concept.

Software developers get their dopamine kick from closing tickets and delivering software. When a software developer holds on to a ticket for a long time, they may become anxious and begin asking: Do others think I am taking too long to work on this ticket? Am I on the wrong track toward a solution? Do they think I am milking time to avoid working on more tasks? When tickets are broken down into smaller units of work, the software developer and the team will see frequent code check-ins and see that their effort is appreciated.

When my team started our shift toward the bite-sized approach, we obtained our initial buy-in from the software developers by showing them how they could work on small bits of the bigger picture and could work simultaneously with others on the team. We made just a few small changes at a time. The initial change was to ask the team to give all tickets a set of subtasks. We then took the subtasks and made them into stories. Each story was then evaluated by the team to identify anything that was unclear and how it could be made clear. We required that each story was given enough detail that even our junior software developers could work on them without oversight. Once the team was used to writing smaller stories with detailed descriptions, we asked the team to figure out which tasks were dependent upon other tasks and which tasks could be worked on without waiting for other work to be completed. After several iterations of these exercises, the team was able to author smaller stories, properly link them together and work on any ticket in the sprint, regardless of their comfort level.

Software developers are required to make a lot of changes when updating their software delivery process. Simply telling them to "take a ticket and check it in when you are done" is not enough to help the team adopt a new process. You will need to implement specific process changes and new methodologies to assist them. If your software developers do not have curious minds and a willingness to learn, they may find it difficult to work together on a new workflow.

Here are common problems that may come from the software developers and some sample solutions.

Problem Statement	Solution Response
I do not like change.	I agree with you, and I feel the same way. Therefore, we have put a lot of thought into a way of implementing a new software development process without jeopardizing our ability to develop software smoothly.
We have been doing fine so far.	Our developers are not giving proper updates on tasks and our workflow has created bottlenecks in the product owner's ability to properly feed updates to our clients. This change was not merely a desire to experiment.
Will this speed up delivery?	We are going to see results sooner and it will help us pivot when necessary, before too much time has been invested in a task. If you recall, we have had to change our implementation a few times in the past because a client's request came in after the work was finished. This new process should help us to catch these issues ahead of time and work around them.
Can we go back to the old way if this does not work out?	We probably would not want to. The tickets will be broken down small enough that the work will make more sense to the software developer. The last thing we want is to go back to a monolithic ticket-writing style.

Documentation Saves Teams

One of the key tools for ensuring a smooth transition and broad acceptance of the bite-sized win mindset is to provide documentation to all key players about your team's progress. It goes without saying that nobody likes paperwork, whether due to the time it takes to write, the time it takes to read, or the impact that it has on the speed of delivery. Often, the business and the client will claim that they love to receive paperwork from the team regarding completed work, but odds are they never touch it. I think we have all been in a situation where we were required to write some sort of documentation to accompany our code, and I equally am sure that, after years of neglecting the documentation and allowing it to become outdated, you found out that no one read it in the first place (which is why nobody realized that it was outdated).

Let's explore a few types of documentation and how they can help you and your team to be successful in modernizing your processes.

Functionality documentation

Functionality documentation is your basic *user guide* for how to use the system. This documentation is written in such a manner that any non-technical user can understand how your software works and how they can navigate the system successfully. Some organizations hire a technical writer for this documentation instead of allowing a product owner to author it, since the product owner may have advanced, non-beginner-friendly knowledge that could bleed into the user guide. This type of documentation is fantastic for your hands-off client because they have no desire to understand how everything works *under the hood* and they just want information about how to use the features that they requested.

Technical documentation

Technical documentation is typically your team's knowledgebase. This documentation provides information about how your software works, how your architecture is set up and any other bits of important knowledge that will either guide a new member of your team or refresh the memory of someone who has been on your team for a while but has forgotten how a process worked. One area that is often neglected is the technical documentation on how your team operates, which would help all the key players to understand what changes have been made and where they can be seen.

Tickets as documentation

When authoring tickets with the bite-sized win mindset at the forefront of your process, you and your team may find that a lot of the work that has been done upfront can now allow your tickets and requirements to serve as your team's documentation. At first, taking extra time to break down tickets into bite sizes will increase lead time for starting your tickets; that should not be a shock to anybody. But as you get used to the process, the difference is easily made up!

Often, the product owner will rush out a lengthy initial document, and you and the software development team must work with the product owner to clarify information and get details wherever needed. Some teams will accept a requirements document as a piece of technical documentation, but other teams prefer to break the requirements down into tickets to paint a better picture of the work that was actually performed. A lengthy technical requirements document does not typically contain code-level implementation instructions, so it is not a prime candidate for technical documentation; the requirements are typically a hybrid of functional and technical information, which is mixing the expectations of the two conflicting audiences (users and software developers). When creating the bite-sized tickets, the aim is for each to have full details that clarify the requirements on a technical level. Some ticket systems also allow you to check in code with the ticket referenced, so those check-in notes will serve as a guide to the final implementation. The combination of ticket and code will later help your software developers to paint a clear picture of what the implementation was and also prevents confusion when there was a difference from the initial requirements. Often times, the implementation does not completely match the initial requirements, whether due to system limitations or external conversations, so in the event that the requirements documentation was never updated, it is safer for a software developer to read code and reference their associated tasks instead.

When documenting under the bite-sized win mindset, you and your team should strive to do a lot of the work upfront by allowing your tickets and requirements to serve as the documentation. Technical documentation tends to be quite specific, so I like to use tickets for this. In Chapter 3: Tasking for the Bite, we discussed epic tickets, which often include the original requirements documentation, as well as any architecture charts and graphs that will help software developers better understand the usage, and so these tickets are often very useful for technical documentation. Functionality documentation is higher level and must be user friendly, so it cannot be presented as a set of tickets, but they can often be loosely based off the product owner's requirements documentation. While the team is learning how to work together to produce documentation, they will learn the value of failing together, which we will cover in Chapter 7: Winning (and Losing) as a Team.

When my team started to implement more of a group effort for creating tickets, we took a day or two to write up all our tickets for the upcoming sprint. The business had initially viewed this practice as a waste of time, because the perception was that we were wasting company time on extended meetings. However, we continued to work this way and did not pay heed to the scrutiny from the business. We found that once the tickets had all been created, we had enough detail on every ticket that any developer could take the tickets and successfully complete the requested work without requiring more information. A positive side effect of this increased time spent on creating tickets was that we could use them later for documentation, saving us yet more time. When our product owner was curious about how the final implementation differed from the original requirements document, we had a trail of tickets that showed the progress and how decisions about changes were made.

A word of warning

When I started working in the education industry a few years back, I wrote a piece of seemingly important documentation on our knowledgebase that was excruciatingly detailed. Figuring that my target audience was a large group of software developers who have a passion for education, I thought this would be a great opportunity to write for an audience who would appreciate my amazing technical writing skills. I sent out a link and asked for feedback. I think this is a good time to mention that not only am I a nerdy guy who enjoys technical writing, but I am also a very sneaky nerdy guy who loves to see tangible metrics. I mention this because there is one minor detail that I never told anyone: I enabled the view counter on the page. In my defense, I wanted to know how much traction I was gaining on the documentation because I had to validate that continuing down the path of writing up documentation was the right way to go. I waited about a month and the counter only increased by two, which I was convinced were visits by my boss and her boss, our director; I did not think either of them even read it, but that was not a concern of mine. This little experiment told me that everybody values *the idea* of documentation, but nobody values the documentation itself. I concluded that gaining an audience in an educational institution is equally as difficult as gaining an audience in smaller organizations who value time saved over quality. You cannot force anyone to read your documentation and, even when the setting is perfect for gaining appreciation for your documentation, it is likely that people still will not care.

Take a moment to think about the old saying, "the chain is only as strong as its weakest link." When training your software development team, make sure to take the time to teach them how to properly write tickets, especially with junior members. If the ticket is clear enough for an inexperienced member of the team to complete, it is clear enough to be an acceptable body of work. Provide help, if necessary, but the goal is to see that they can complete the work with minimal intervention. The team can come together and relish the fact that they have been able to author tickets with such a level of detail that they were able to give an opportunity to someone who may not have had such an opportunity in the past. Senior software developers will be happy that they may not be pigeonholed anymore. Every software developer has an area within the software that they do not want to work in. Forcing a software developer to work on something that they view as mind-numbing will have a negative impact on software delivery. Whether we want to admit it or not, when we are forced to work in an area that we dislike, we tend to take longer to deliver the work. When your team can cross-train on systems and allow everybody to share equally in the work, and when you break down tasks small enough to where software developers are not confused about how to do the work, you are going to find that the work is done more efficiently. More people will buy into the process as the results speak for themselves.

Summary

When your project's key players have a positive experience, it means that your team is doing something right. As we have discussed, the goal of the bite-sized win mindset is to keep everybody happy during the development and delivery process for your project, whether your team delivers the project faster or not. The perception of efficiency and progress generated by delivering and preparing work in consistent intervals is often enough to turn a negative opinion into a positive experience. It is also a great way to turn a mundane delivery process into one that keeps everyone talking.

THE BITE-SIZED WIN MINDSET STRIVES TO MAKE EVERYONE INVOLVED IN THE PROCESS HAPPY

BY HIGHLIGHTING PROGRESS AND CREATING AN AVENUE FOR INTERVENTION BEFORE THE FINAL DELIVERY

Chapter 6 | Separation of Concerns

Separation of concerns (SoC), as a concept, is hardly new. In fact, Google Books Ngram Viewer[2], which measures word usage over time, shows the first usage of the term back in 1896 with a huge spike around 2006.

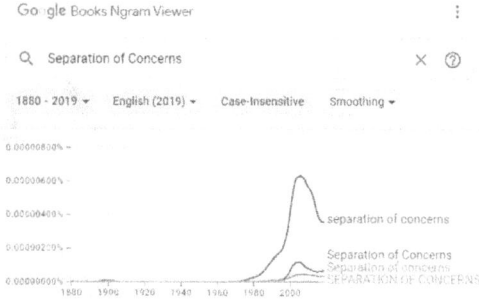

It is not even that new in computer science: I searched high and low, and the earliest appearance of the term *separation of concerns* in relation to computer science that I could find was in a 1976 paper by Edsger W. Dijkstra, titled *"On the role of scientific thought"*.[3] The Wikipedia article on separation of concerns further references Chris Reade's 1989 book *Elements of Functional Programming* as eventually defining the term for use in software development. In 2003, Robert C. Martin (also known as *Uncle Bob*) wrote a book titled *Agile Software Development* where he first introduced *SOLID design principles*. SOLID is now an industry standard for object-oriented programming, and one of the concepts (namely, the S in SOLID) is the *single-responsibility principle*, which

---

[2] https://books.google.com/ngrams/

[3] https://www.cs.utexas.edu/users/EWD/transcriptions/EWD04xx/EWD447.html

is another acceptable term for *separation of concerns*. Many consider the introduction of SOLID to be one of the first widespread adoptions of separation of concerns.

In software development terms, separation of concerns is the practice of preventing the co-mingling of competing logical functions within a single logical function. My definition is, admittedly, a nightmare to read, so more simply: *Things should only do what they say they are doing.* Typically, software developers will be explicit when naming their methods. For example, if a method is supposed to add two numbers together, it will be named something like *AddNumbers.* This method should not additionally print out the answer at the end, since that is a different logical functionality.

We can also apply the separation of concerns approach to writing tasks, to ensure that the work you have assigned to a software developer does not include functionality that competes with other tasks. We should strive to write tickets that are concerned only with a very specific function within the software. This practice is very important for the bite-sized win mindset.

I have found that younger software developers tend to have a hard time figuring out how much functionality is *too much* within their methods. When confronted with this confusion, I like to reflect on when we initially learned how to program. Often, a future software developer's very first program will be the tried-and-true *Hello World* program: a program that simply prints out, "Hello, World!" This is simple to accomplish and does exactly what it says that it will do and no more. This example is their first exposure to separation of concerns, although they are likely not taught about this concept, yet. Next, they will likely be tasked with building a program that accepts two numbers as input, adds them together, then prints the results to the screen. Once again, the program does exactly what it says. Soon after, it will be time to learn about functions and methods. The addition program will need to separate out the addition functionality from the printing functionality. We will call this the *AddNumbers* function. The *AddNumbers* function should add the numbers and return the value, and then the rest of the program can handle printing the output. Students who do not naturally see why functionality should be separated will often print the results *within* the *AddNumbers* function. It becomes imperative to impress the importance of ensuring competing logic is not co-mingled in your methods. Adding and printing do not have the same functionality.

Architecture patterns can isolate concerns

Separation of concerns is used throughout many areas of software development. For example, it used to be good practice to isolate work by separating the user interface from the backend and the backend from the database. This is your typical N-tier architecture, an architecture model that has been used as far back as the early 1990's, when web application systems were popularized. We will look at this architecture along with the microservices architecture, as they exemplify the separation of concerns methodology.

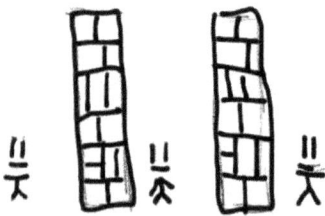

N-tier architecture

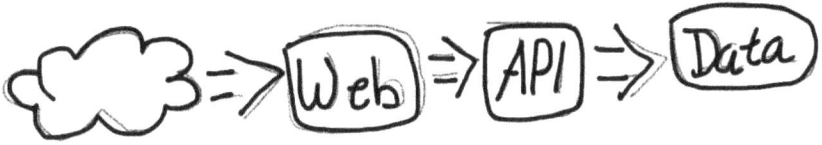

N-tier architecture is a software design pattern that serves as a great illustration of separation of concerns, because each software layer has its own isolated project. You typically have a single codebase that contains three separate projects: database, backend code (such as the main API logic), and frontend presentation code. This code separation made it easy for a software developer to figure out which project housed which code when they needed to make changes, since the logic was held in only one of three projects.

The risk in this model is that, if a software developer is unsure of which layer to house their logic in, they might put it in the wrong place, so you can wind up with weird logic sitting in a layer that it did not make sense for; this makes the logic difficult to find and update when necessary. To mitigate this, teams will often be more diligent about enforcing where code should be located during their peer code reviews.

When I first began programming back in the early 2000's, this was considered the de facto way to organize a website's code architecture, especially in smaller organizations that needed a tried-and-true software architecture structure without any special requirements. Gradually, this practice was incorporated into the building blocks of more formalized architecture structures, such as MVC all the way up to Microservices. The MVC (Model-View-

Controller) design pattern uses elements of N-Tier and serves as a great way to separate areas of code, though some software developers consider it to be the same model.

Microservices architecture

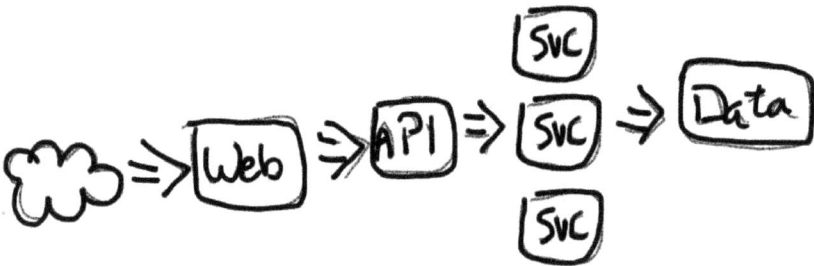

In modern architectures, code is separated a little differently. We have mostly moved away from N-tier architecture, and we have begun working instead with specialized technologies that will separate out our concerns, such as *microservices*. Microservices are more of an architectural design pattern as opposed to a software design pattern. A microservice is an application service that serves a single purpose, intended to be able to respond to a request in an exceptionally short amount of time. For example, one microservice may be responsible only for retrieving data from one database table and it would be expected to do that task in a matter of milliseconds. The rise of microservices enables teams to move entire modules into their own code repositories and allows developers to separate themselves from monolithic project architectures that could contain millions of lines of code in one place, to focus instead on one module in one repository.

I once met a software developer who was working on a team that had hundreds of microservice projects throughout their infrastructure. Each microservice had one expected input and one expected output. Each microservice was expected to respond within a fraction of a microsecond. Achieving those sorts of times is impressive, and it shows how efficient our code can be when it is only expected to do one thing. This is how I want you to think about your ticket structure.

Creating tickets with SoC in mind

Creating a story is not difficult. Creating a story with a reduced scope *can be* difficult. When combining separation of concerns with ticket-creation, you need to think about what the requested unit of work is going to produce; each ticket should result in something tangible added to the project. However, there is no guide that tells you how small your tickets should be, and you will likely have to define what constitutes an appropriate amount of work for a ticket for your particular team or project.

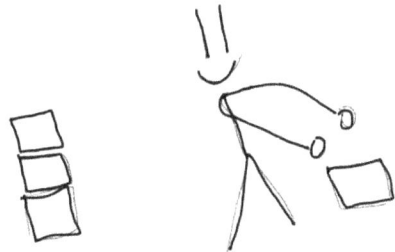

For example: In my team's delivery agreement, we have specified that the smallest possible unit of work will always deliver something that can be viewed or used by a client without creating an issue for the system's existing usability. When creating new tasks, then, you want to ask yourself, "What is the smallest amount of work that can be produced that has a usable, non-intrusive deliverable at the end?" Keep in mind that a *deliverable* does not necessarily mean the entire feature. Providing a deliverable simply means that when the code is checked into the repository and deployed to a client's environment, it can be used by either the client or another existing process. A deliverable can even lack function but should not negatively impact your client's experience. *Why would we deliver something that lacks functionality?* The deliverable can be viewed, and it does not create an issue for the system's existing usability, so is still in line with our team's delivery agreement.

Your deliverable should not be half-baked, unsightly, or cause concern in the client. This means that you can deliver a UI element that does nothing, but you cannot deliver a UI element that causes your user to believe that the system is broken. You can also deliver code that moves data around, but if it causes something negative to happen, such as breaking a credit card integration subroutine, it should not be in your codebase.

The question of how small a deliverable should be does not have a universal answer. Nobody can give you a textbook definition of the size of your requirements; this is something you will have to work out with your team, perhaps on a project-by-project basis. Of course, there are best practices, and your team can agree upon some high-level criteria that might define the size of a task. You should also look at your existing work and think about the impact that new requirements may have on the team's ability to deliver on requests. Specifically, if you have a task already in the pipeline that is scoped for the same section of software that has a change request, you may have to ensure that your new request does not conflict with the requirements that were already vetted by the team. You also need to consider how large your existing work is, in order to prevent multiple software developers from altering the same section within your software at the same time; this causes conflicts in code, which are not always easy to reconcile.

To decide on the size of a task, ask yourself:

- What tasks currently exist at the time of the request?
- Are you able to utilize the code after it is checked in?
- What other functionality does this task affect?
- Is this task enhancing existing functionality?
- Is this task introducing new functionality?

The product owner should assess whether the task being requested interferes with existing logic in the client's system. If your task may cause existing logic to break before subsequent tasks have been delivered, you probably do not have enough functionality defined in the story, or your request has been broken down into too many tasks. However, if the task interferes with existing logic, but does not break the existing logic, then it is likely that your ticket has successfully isolated the concern and can be considered a properly-sized task.

As mentioned, it is also acceptable to deliver code that does nothing. You might deliver a button on the page that does absolutely nothing and use a feature gate to hide it when it is not necessary to display. You might submit code that will only be activated once the complete feature has been finished. These are acceptable because they are still working toward completion of the overall goals. Just because code is in the codebase, does not mean that it needs to be used. The tiny wins that move toward the end product will serve as a status update for all. You should, though, be wary of allowing dormant code into your codebase. *Dormant code* (often called zombie code) is code that is never referenced or utilized. For example, you may have a function that calls a new API, but the API is abandoned before the project can be finished. The function is never used again, so we consider that dormant code and should strive to remove it. When you are dealing with separation of concerns, this can occur if a newer task negates the task that was supposed to create the API, so you and your team should regularly scan the code for unreferenced, dormant code.

As an example of how you might break down a ticket to ensure separation of concerns: Say you want to introduce new logic that allows a user to change or delete their profile picture. If a task is written that allows a user to delete their existing picture before you introduce the task that allows them to update it, you may inadvertently cause a workflow issue for users who want to keep a profile picture. A *workflow issue* is an issue that is introduced into your software when a user expects to perform an action, but they are unable to complete the action because part of the functionality does not exist in your software yet, or another function of your website interferes with their ability to complete their expected action. In this scenario, you may want to reconsider how the task was separated and prioritized, since you have introduced code that is going to negatively impact a user's experience. Instead, you could have a predecessor ticket that enables or disables the delete functionality via feature-gating, as we had discussed in Chapter 3: Tasking for the Bite. Alternatively, you could just have the *update* task set as the predecessor for the *delete* task.

Summary

The point of the bite-sized ticket is to think about the immediate concern and what the ticket hopes to achieve. If you can deliver work that deals with that concern and prevent tragedy from striking when that work is deployed, then you have successfully separated a concern.

SEPARATION OF
CONCERNS IS THE
PRACTICE OF
PREVENTING THE
CO-MINGLING OF
COMPETING
LOGICAL
FUNCTIONS
**WITHIN A SINGLE
LOGICAL FUNCTION**

Chapter 7 | Winning (and Losing) as a Team

Your team is used to delivering outstanding software and keeping a smile on your client's face. Your team is also used to creating bug tickets and attempting to shift blame onto someone else for causing the bugs. In this chapter, let's explore how our teams can benefit from not only recognizing and celebrating their wins, but also celebrating lessons learned from their losses. And, importantly, doing so as a team, and not shifting blame. While we may all want to believe that we have no faults, it is important to remain humble, share our successes, and recognize that we can turn a loss into a win for the whole team.

Honoring wins and losses

Imagine, if you will, that you are being interviewed for an amazing opportunity with a great company. After an outstanding

conversation, the interviewer looks across the table at you and says, "Tell me about a time when you succeeded."

I am certain that you already have some stellar stories just screaming to be called upon to help you shine in this moment of need. I am sure that you have had so many great successes throughout your career that choosing just one accomplishment would be nothing short of the toughest decision of your life. You thoroughly prepared for this interview and are ready to tell the interviewer about the time that your company turned a profit in excess of one million dollars overnight, just because you authored a fancy new widget on the homepage. From problem identification, all the way through to the final delivery of this widget, you were at the helm. This achievement paid homage to every experience that you have had in your career, and you orchestrated its development in such a way that even Dennis Ritchie, inventor of the Unix operating system and the C language, was said to have shed a tear at how beautiful your widget was.

After you finish your moment of bragging and recompose yourself, the interviewer then says, "Okay, great. Now, tell me about a time that you failed."

If you are anything like most of the people I have interviewed throughout my career, you are now staring ahead like a deer in headlights because you did not prepare *that* story. I have conducted many interviews and when I ask this question, I typically get a response along the lines of, "I cannot think of any failures because my career is full of so many successes!" You may view this as a fair response, since it shows that you are confident in your experience, but in reality, interviewers ask this question for a reason. The interviewer is not necessarily looking to knock you down a peg or for you to humble yourself. The interviewer is interested in knowing whether you can learn from prior mistakes or if you are apt to repeat them. Learning from your mistakes is a very valuable skill that a lot of people seem to lack. I like to look at successes and failures both as learning opportunities, regardless of which end of the spectrum they are on.

I am not ashamed to admit when I have been wrong and when one of my decisions did not work out. In fact, one of my greatest successes was nearly one of my greatest failures. Many years ago, I worked with my team to rewrite a payment processing algorithm. We spent months on this algorithm and did loads of testing to ensure that it would not fail. There was a lot riding on this payment processor and anything other than success was out of the question. We eventually finished writing and testing our code, then deployed it to production.

The first time one of our clients used it, we discovered we accidentally caused a loop that would duplicate payment details. Luckily, those payments were not submitted. Our failure could have been disastrous, but instead we learned from everything that we did incorrectly and made major adjustments not only to our code, but also our way of thinking. By making these changes, we turned this failure into our greatest success: we took a payment process that took about an hour to execute and whittled it down to about 45 seconds. This was not a common feat to achieve, and it was such a success that I still love to brag about it today. It goes without saying that I love to brag about our failures as well.

In this chapter, we will explore a little more about winning and losing as a team and why you should celebrate both.

Celebrating the wins

A team's success is not only measured by the products that they produce; a team's success is also measured by the lessons that they learn. A team's ability to identify what caused them to succeed is just as important as their ability to identify what caused them to fail. Learning from failures can grow the team's bond.

When your team has a win, be it anything as small as a hotfix delivery or as large as completing an entire feature, it is important to celebrate regularly to remind the team of their wins and how they came to achieve them. This does not always have to be a celebration with cake and confetti, but simply giving kudos to the team for a job well-done will do wonders. Make sure everybody understands the role that every member played in delivering the win and use this opportunity to increase general morale. It is common for teams to get into a cadence of delivering work and shipping it, without taking the time to appreciate that they delivered a win. The team grows more effectively when they know which direction to grow in.

When your team has completed their implementation of the bite-sized win mindset, you should take the time to celebrate. The mini celebration will assure the team that their efforts resulted in a positive experience and encourage them to continue the practice. A bite-sized win deserves a bite-sized celebration. At first, your team may find it a bit odd that you are encouraging them to celebrate simply doing their job well, but when they begin working on new tasks, they will think back to the team's celebration about finishing the implementation and that the whole team agreed on the success that the bite-sized win mindset was the appropriate approach to creating tickets and tackling work.

Perhaps you are lucky enough to have had a manager who was attentive, supportive, and who gave you kudos on a regular basis, maybe complimenting you for doing work above your typical skill level. By receiving constant, reassuring acknowledgments, you can see that you are on the right path. With these experiences, you are more able to shape your career and focus your energies on how you would like to conduct yourself professionally, which not only benefits you, but also the team you are part of. These are the conditions we want to inspire with recognizing bite-sized wins.

Sharing the win

When the team experiences a success, the contributions of all members should be acknowledged. If an individual stands out, it is okay to give them their own individual praise, but do not neglect the team as a whole. As time goes on, the hope is that a successful individual recognizes the efforts of the team as part of their success. It is not fair to celebrate a team member's success and end it with a quip like, "Do not forget that we were here to support you." You may inadvertently shatter their ego instead of teaching that the team wins as a single unit. Instead, set the example of giving the team proper credit whenever *you* are particularly successful, and you may find that they follow suit in their own successes. We all like to receive praise when we do something good, but it feels just as nice to share that praise with others.

When I deliver a major hotfix (software patch) to a client, I do not take full credit for the hotfix, because I understand that I am not solely responsible for the delivery. Yes, I wrote the code. However, a product owner engaged the client and discovered the original concern. When the fix was prepared, another software developer reviewed my code. Yet more software developers approved my pull request and QA tested my code. Maybe I consulted with a few software developers on possible issues or solutions. My manager was there to support me and ensure that I had all required resources available. Lest we forget our Scrum Master, who ensured that nobody violated the sanctity of my time. At the daily stand-up meetings, I was able to consult the team and let them know my status; the team also kept me honest about my time. If the mechanism for deploying code to the client server is an automated process, then somebody had to set up the initial automation and someone else out there is maintaining it. The fact that I wrote a piece of code does not mean that I was the only one involved in the success.

When I am a member of a team, I reject the notion that I succeed on my own. When something good happens, I always make sure my team knows we all played a role in it. Even if we come across a bug that only I apparently have experience to fix, it does not make sense for me to celebrate that win on my own because others had to look at the issue before I could succeed. The inverse is also true: perhaps there was something I failed at that someone else was able to learn from, using it to formulate a new plan to resolve an issue in the correct way, or at least a different way.

So, I reiterate that you should not celebrate only yourself for succeeding. You should always thank your team for creating the avenue down which you could succeed. But more than that, you should also celebrate (to some extent) when you *do not* succeed.

Celebrating the losses

It may seem counterintuitive to celebrate the fact that you and your team have failed. It is natural to be happy and celebrate every time you win, but the fact is we learn more from failures than we learn from successes. Failures, therefore, are also worth paying attention to. I will give two examples of why it is important to focus on failures.

First, we tend to recall a failure when we find ourselves in the same situation again and try our hardest to avoid repeating the same mistakes. That is the hope and the point of recognizing your failures. We recall our failures because we do not want to fail again. Memory can be a funny thing; when we fear a failure, our minds will often serve up a memory of the last time that we failed in a similar scenario, even for a long, long time after the initial failure. In my experience, when I eventually forget my own failures, it is often because the failure has become a lesson learned and I no longer need the memory of what initially caused the lesson. I believe that the brain does not want to clutter its memories with negative thoughts and experiences; we should strive to populate our memories with solutions to problems that we have faced, as opposed to failures that we still need to recall.

Second, people tend to remember when somebody *else* did something wrong, but only if it was recent. Think about the news, both local and national. Emphasis is typically on negative stories, especially when it comes to politics. When a politician is up for reelection, most of the news stories will typically be about their failures while in office, regardless of how many successes they may have achieved. *Why* is a mystery to me, but people like to remind each other about things that others did incorrectly. However, I have a secret for you, which may help ease your mind: people around you will forget your failures over time and people who want to remind others about your failures will not always have the energy to do so. Even when those failures may be downright funny and at first everyone is discussing them daily, those memories fade as time finds a new victim. During the election cycle, we are constantly reminded about politicians' failures because political ads are relentless. After the election cycle has ended, we forget about most of the failures because nobody is talking about them anymore and most of the failures likely did not impact us directly. Knowing how easy it is to forget the failures of someone else, you should not dwell on your own failures beyond their ability to teach you things; nobody else is.

It is important to hold yourself accountable when you do something incorrectly, and to be aware of any mistakes your team makes. If a coworker decides to be cheeky and recalls your team's failure in front of the CEO, be prepared to counter with the success that the team was able to gain from that failure. Undoubtedly, any failure has some sort of negative effect, but make sure that you and your team are able to appropriately react to those negative effects. You can always regroup and start over again. What you cannot do is "relearn" a lesson that you never learned in the first place. Keep the successes at the front of your mind and allow the failures to become lessons learned.

Take the time to pick apart your wins but take even more time to pick apart your failures. I would not suggest orchestrating a party for a failure, nor do I expect you to experience happiness, but do raise awareness to the fact that it happened and make sure there is a lesson to be learned. The failure should remain in your memory and your team should be able to discuss it freely without feeling shame or blame.

Once a failure has become a win, the team has learned a valuable lesson that cannot be taught with words alone. We learn best from personal experiences. The team should be able to recall when they failed, why they failed, and how they overcame the failure to have open dialogue in future projects. The team can then apply that information to new projects to make sure the failure does not happen again. We should strive to always be looking for new ways to apply our experience to future situations and overcoming failure.

In short, your losses can easily turn into wins. You are not celebrating the fact that you failed; you are celebrating the win that can or did come out of the failure. You are learning a lesson, applying it to future projects, and ensuring that the failure does not hold you back or repeats itself in the future.

Types of failures

There are different types of failures, and each will teach us different lessons. For example, lack of knowledge is one reason for failure. It is often difficult to know what to learn from these losses since, ahead of time, you do not know what you do not know. In this case, I would suggest you keep a log about how failures have happened and how the lack of knowledge contributed to failures, and work with the team to formulate a plan to provide everyone with more knowledge around any area that nobody seems to be familiar with. This is typically discussed in your team's *retrospective meeting*, where everybody gathers and reflects upon how either a project, sprint or incident resolution went; the point is to highlight all the successes and point out any failures and how to avoid them in the future.

A *slow-release failure* is another type: this is a failure that has not yet caused all the damage it can do, but over time, you will begin to see the effects. For example, if your code is calling a routine that edits data in an unintended manner, it may negatively affect your data significantly before anybody notices and is likely to cause even more damage as it goes unaddressed. When we recognize that a slow-release failure is occurring, we should take note of the events that led up to it and the events that occurred after it. This information can be more valuable than identifying the initial failure itself because it gives you data on what signs to look for in the future. You can then apply those signs to other scenarios and figure out if you are potentially at risk of failing elsewhere.

Recognizing that you are on the verge of failing is a valuable skill. If you possess the foresight to recognize that a failure is enroute, you will be a great contributor to your team's success. Taking the time to recognize the signs before and after failure is important to help you and your team succeed in the long run. As some practical steps, I encourage you to build up a list of common patterns that will help your team to identify future failures. Data-related slow-release failures are sometimes easy to catch because users will notice that something is *off* with the data that they are examining. If you have a financial system, you can often catch these slow-release failures by testing simple calculations. If your system has built-in safeguards for bad data, your system will be more resilient to slow-release data failures. As your team builds up an intuition for where and when failures may occur, they can add in tools and workflows for handling these potential failures before they occur.

When to avoid celebrations

I once posted a short blurb on my LinkedIn feed about winning and losing as a team, which mentioned that you should not be afraid to celebrate your losses. In response, I received a comment that was interesting, although maybe not articulated in the best way. The commenter was unhappy about the fact that I said a team should celebrate losses because they were under the impression that there is a current revolution in the industry that encourages teams to be sloppy and receive kudos for doing things incorrectly. I thought about it for a while, and I concluded that this commenter had not considered that I was referring to the usage of failures as a learning opportunity. I do believe that their underlying concern was that you should not always look at a loss as a positive thing. I agree that not all losses are a cause for celebration, nor do I believe that a culture of religiously celebrating failures is a good culture to cultivate. There are times when the team should learn from their mistakes without overtly celebrating the negative consequences.

I learned from the reaction to my post, and I feel that my point about celebrating losses may require some clarification to avoid any future confusion. I want to make it clear that there will be times when you will not want to celebrate the loss. Celebrating a failure is fine when, say, just before the launch of a new software feature, the team realizes that some old code is being executed incorrectly and can patch the code so it no longer calls that bit of unwanted code before it can do any harm. Celebrating a failure is *not* fine when the team launches a new feature with these mistakes in place without anybody knowing and the company loses the equivalent of six months of your salary within ten minutes. You might want to celebrate once you overcame the loss; celebrate the lessons learned and move on.

It is also not appropriate to celebrate repeat failures. For example, say that your team was implementing a new technology into the core codebase but had failed to properly utilize its basic features. Perhaps they deployed the code to the production server and, at the end of the first day, nothing worked anymore. The issue was so bad that the team had to roll back all their code and release an old version while an investigation was underway. After the initial failure, the team figured out the bugs and found a way to make the new technology work. This moment, when the fix is found, is a good time to celebrate the fact that your team overcame the hurdle. Then, say that, months later, the team implemented this new technology in another section in the code and inadvertently repeated that same failure as the first time, causing the same code rollback to happen. It seems that your team did not learn any lessons. This time, even once the fix is applied, celebration may not be appropriate because the team is repeating the clean-up work instead of learning new solutions from the failures and formulating plans on how to not repeat them. Overcoming a repeated failure is not grounds for celebration, as it is not exactly a new win; it is just a recycled win.

Dealing with repeat failures

We can either learn from failures or be reprimanded for repeating them. If no one actually learned the lesson, it is a reprimanding scenario. However, do not ever be visibly upset with your team when the same failures keep happening. Instead, recognize that the issue is going to continue happening until you formulate a plan to overcome the obstacle. Teams without plans are like lamps without light bulbs – they serve a purpose, but there is something missing.

Investigate why the same issues are repeated. There could be some information or process ingrained in the team which causes them to repeat actions even though these actions have led them wrong in the past. Identify the cause for the repeated failures, and then strive to anticipate and plan for them.

There comes a point at which you need to stop accepting the failures and start putting better plans in place to mitigate them. The discovery phase is a great time to implement these plans. In this phase, before tickets are created, the software developers discuss their experiences related to any upcoming tickets based on known requirements and recall any issues that arose the last time similar work was performed, including failures. If nobody can remember what these issues were or how they overcame them in the initial discovery meeting, then the team needs to come back to the ticket when all the information has been prepared. It is not appropriate to accept a ticket that has known issues, but no reference to those issues.

Preparing for unknowns

I have worked with some massive software systems and there were sometimes cases when, after an issue was fixed in one area, an unexpected breakage occurred in a completely unrelated area. This usually comes from poor code architecture or insane code complexity. In this scenario, it is common to find that the software developers do not understand the code as well as they believe they do. When faced with the recurring theme of fragile software breaking, your team will want to identify which areas exhibit the most common breakage scenarios. There may come a time at which the team will need to evaluate the current implementation of a section of software and possibly replace it entirely; although it seems like a burden to do, the team should be willing to accept that desperate times call for desperate measures. In recent years, I have emphasized that when my team accepts work, we must consider the overall complexity of the possible implementation and recall any experiences with altering anything in that section of the code.

Another tactic for dealing with unexpected software breakages is to always give ourselves extra time for inevitable *unknowns*. Specifically, if we are aware that the element of risk could cause the implementation to take longer than expected, we will give those tickets higher points than we might otherwise. It is very difficult to predict the amount of time that an *unknown* will take to resolve, but you will likely develop a general feel for how much extra time to pad into your schedule. The bite-sized win mindset helps reduce the amount of time it takes to deliver extra work since the tasks are already bite-sized and easy to navigate around.

There will be times when the code is so complex and disorganized that there will inevitably be issues with it after the software development phase has been completed. In these cases, everything tends to look fine immediately after the work is completed, but days later, a seemingly unrelated section of the software will be found to be broken. Perhaps the code is known to be unstable, so to counter this kind of issue before it occurs, your team could start to incorporate additional testing scenarios before code is committed to the repository. You can also set expectations for the business by letting them know that once the code has moved to the QA phase, they should expect that it will not be finished for a while afterward. The development phase may take hours to complete, while the QA phase may take days. Try to jump in front of the problems when you can but acknowledge when an unforeseen problem has occurred. We all want to be able to give an exact estimation for how much time a task will take to complete, but if an unknown circumstance arises, do not be afraid to put your hand up and let others know that it is taking longer than expected.

It is very difficult to debug something that is massive. It is very easy to debug something that is very small and pinpointed. When your team implements the bite-sized approach and stops working with large, complex sets of requirements, you will find that team failures have far less impact than before. Sure, you will still experience losses, but your team's ability to organize around them will be swift and effective. So, go forth and enjoy those mini celebrations but do not be ashamed to admit when you need to focus on the losses.

Accidental wins

Sometimes, you will simply succeed due to dumb luck. You should pay special attention to wins that were due to dumb luck because it is very possible that similar scenarios may come up in the future, so your team will want to build a strategy to ensure that a failure does not happen the next time dumb luck is not on their side. There is not a lot to learn when luck comes into play, because the win was not planned for. Ask yourself how many times you have found that your code works, and you have no idea why. Did you return to the code and figure out why it works, or did you throw your hands in the air and leave a comment that it should never be touched again? Unfortunately, "it just works," are not words that help a team grow. Always investigate when something is working, and you do not know why.

A win is not always due to incredible experience and skills, but sometimes happens by accident. Take the time to understand why you succeeded, because in the future that particular set of circumstances may not be there. Your team should strive for purposeful wins. I would suggest that your team's delivery agreement should state that code should never be committed to the main repository if the software developer who authored it is unsure of what it does. Sometimes a software developer will copy-and-paste code that they found online and fail to understand the context around its full use. Other times, they included a math calculation that seems to work for their scenario, but do not investigate other scenarios that will try to utilize the calculation and give odd results. You want wins that the team has honestly achieved, as those wins are reflective of their knowledge and strength. Dumb luck is not strength; it is dumb luck.

Achieving a major win by pure luck is not cause for celebration. Of course, for any win you should quietly pat yourselves on the back, but recognize that luck is not always going to be on your side. I do not mean to sound pessimistic, but luck is temporary. Luck is not constant. If you did not study for an exam but luck allowed you to pass anyway, you did not earn that exam's achievement. You did not learn the information that was required to pass the exam. Yes, you achieved that passing grade, but if you were to be re-examined in the future, luck may not be on your side any longer.

Summary

A win and a loss clearly have different outcomes for the state of your software, but they share a similar trait in that they can both be used as lessons for everyone. Always ensure that you are looking out for the opportunity to learn, be it from the win or the loss. In the end, the lesson that you learned is more valuable than the knowledge you had applied.

A TEAM'S SUCCESS
IS NOT ONLY
MEASURED BY THE
PRODUCTS THAT
THEY PRODUCE
**A TEAM'S SUCCESS
IS ALSO MEASURED
BY THE LESSONS
THAT THEY LEARN**

Chapter 8 | Capitalizing on the Wins

When your team begins to see successes from bite-sized wins, I recommend that you draw attention to the fact that they were able to deliver something large in very small pieces. It is great for morale. It is great for building the confidence of your team. It is great for eliciting trust from the business and your clients.

But beyond just celebrating, you should also find ways to capitalize on your bite-sized wins. Casually highlighting the win in the next company meeting, for example, may be a great way to improve the overall perception of the bite-sized process and help everyone involved to trust the process. Since you are in control of the narrative, you can influence the tone for the company's reception. As we had previously explored in Chapter 5: Working with Key Players, the success of your process-change implementation relies on buy-in from all the key players within your organization; highlighting a valuable win as it happens is one of the ways that you can capitalize on your wins and gain that trust, and that buy-in.

Advertising your wins

Getting the most out of your bite-sized wins may be successfully accomplished by, for example, advertising that your team had a quick, seamless sprint and effortlessly delivered the client's feature in less time than anyone could have hoped for. Without proper advertising, the hands-off client may regard the delivery as normal for the schedule and not make note of it.

The hands-on client may notice that the software delivery is increasing in frequency, but they may be so busy with their remaining tasks that they do not realize the significance. The business may notice the increase in the amount of work that can be discussed with the client at earlier points in the process, but they may not know how to approach the client to showcase the importance of the uptick in work that is ready to review. It is up to you, therefore, to ensure everyone is aware of your successes as they occur.

Imagine that somebody within your organization, regardless of their level of seniority, is reluctant to allow changes to the software development process and has been pushing back on your implementation of the bite-sized win system. As you start to see positive effects from converting to the bite-sized win mindset, make sure you inform them of the status and benefits of the updated processes so they can appreciate the work that has been done, the transparency the process provides, and that modernizing processes is an important step in delivering great software to your clients. If they are unaware of the process change's effects, they may remain unwilling to support any changes and continue to see bite-sized wins as a waste of time. Nobody wants time wasted, but everyone wants to see results. When your team has these results to show off, it is the job of the team's leadership to ensure that the successes are advertised (in a tasteful way). Of course, rubbing someone's wrong assumptions in their face is not the best approach and could result in an inverse reaction. Casually highlighting the positive outcomes is more likely to obtain more buy-in as the efforts progress. This advice holds especially true when the team decides to implement a process change slowly; these direct, purposeful highlights are crucial for progression.

I have run into this scenario myself, wherein my team was making impactful changes to our process and the business did not see value in these changes. In line with the advice in this chapter, we had to find ways to tastefully advertise our wins so that the business would find value in our changes and allow us to continue down our journey to finding the most efficient process possible. Everyone likes to see results, especially if those results raise the company's profile within the market. I can assure you that showing off your results as they occur will make even the most stringent naysayers quick to appreciate and accept the changes.

You will find that there is often still "that guy" in the corner who is very pessimistic about changes to how a software development team works, believing that the team's years of experience means they have already tried everything under the sun to fix their issues. Perhaps they believe that the team is already working optimally and there is nothing left to change with the team. On the flipside, they might even see the team as a lost cause and do not see any benefit in trying to fix it. We want to raise expectations. We want the team to be not only well-oiled, but highly optimized. We cannot optimize a well-oiled machine when there is friction in obtaining buy-in from the business and the client, so it is important that your team is able to successfully advertise their wins so that everyone involved in the process can understand and appreciate the efforts being put forth. Do not forget that "everyone involved" includes your team members.

Team morale

Capitalizing on the wins should not be directed only at external forces. There will be times when your team's successes are not acknowledged by every member of your team. Software developers do not like change. Have you ever watched Family Guy? I was a pretty big fan, back in the day, referencing the show far more often than was appropriate for any conversation. I am about to do so again, so bear with me. There was a scene wherein Peter had destroyed the side of the house by Stewie's bedroom. The camera zoomed in on Stewie shouting, "What's this? There's something wrong with the house! I don't like change!" Every time I see that scene, I think of myself and of other software developers. Software developers are not necessarily fragile beings, but I do believe that it is easy for their morale to shrink when a new process is introduced.

When it comes to the day-to-day activities, they (we) tend to be very robotic in the way that they operate. When introducing a new process, they worry that they will not follow it correctly, or that it will negatively affect their productivity. This notion can be reinforced by the fact that when you change your process, you are likely not going to see immediate results. The hope of the bite-sized win mindset is that you do not disrupt the current workflow too much, but your goal is still to overturn the process, albeit in a positive way. It is therefore important to be aware of your team's morale and what might affect it.

When you convert to the bite-sized win mindset, a team will typically expect to see faster deliveries immediately. After all, they are experiencing more (smaller) wins than usual and the natural progression is that your overall delivery is going to be faster, right? This simply is not always the case, nor should a faster overall delivery be the short-term goal. If you recall from the book's introduction, I had stated: Make no mistake, if you break work down into smaller chunks, you are not guaranteed an instantly faster delivery, but it will often buy your team the full length of time required to finish the software. There are a lot of processes involved in making a delivery happen and to think that a single process change related to tickets is going to somehow cause everything to work exponentially faster is just not realistic. You must make sure to manage your team's expectations to keep up morale.

I suggest that you keep your team abreast of what is happening daily and keep them informed of results as you start to see them come in. Let them know that what they are doing is having a positive effect on the business. If you are doing your job to advertise the wins, as described above, the team's efforts are not going unnoticed and the changes to the software development process have been reflected positively on the team.

To keep up morale during the change in process, capitalize on the attachment your developers have to their work. Like anyone who creates, software developers will have a sense of pride and responsibility for the software they develop. They have invested time and effort in this software, after all. If you are a software developer or have ever worked as one, you are probably very familiar with the great feeling associated with seeing your name in the history of check-ins for a project. Say you wrote a class two years ago that has been fine up until now, then you see a request for a new feature to be added. You see your name in the history file and know that your code was solid enough to run reliably for two years without any alterations. The sense of pride in knowing that your skills are tried-and-true is indescribable. This sense of pride is what you want to invoke when you capitalize on the wins gained by your software developers. Make sure they know that the wins are their doing, and then encourage them to take pride in the fact that they have delivered something so valuable.

Note that this pride can be a double-edged sword: when you change a process, the developers may feel uncertain that they will continue to see the level of results they are accustomed to, so they will likely be a little deterred. Morale is essential, but it is not always the easiest metric to gauge. How do you know that your team is truly happy? One way to judge this is to look at their level of adoption of the new process. If you notice that the team is lax in adhering to the process changes, it may indicate that they are losing their morale. If the amount of work being produced is far lower than expected, it may warrant an investigation: is the workload incorrect or is this an indication that there is an issue with the team's morale? You should know your team and understand their abilities before you upend any of their processes so that you can gauge where they currently stand and predict where they are likely heading.

Keep your team happy. Ensure that they always know what is going on. Ensure that they have insight into the successes that they are generating. Do not allow your developers to work in a bubble. Do not be a black box that prevents your team from seeing the big picture. Seeing their results and knowing how to look for them will naturally evoke a high level of morale in a team. Know that some teams will require a bit more handholding and so requirements for your efforts will be a bit more intense. Make sure your development team knows when they are doing a great job and the morale will follow.

Business perspective

I have mentioned previously that the business is going to be the least likely to buy into the process change. This assumption is based on my personal experience of being on software development teams who worked in tandem with the business. Typically, the business believes the software development team exists for the sole purpose of writing software and shipping it out; as robots who pump out software that suits their needs, and they serve little other purpose. There is typically no regard for the underlying process involved in writing

software, and I believe that is why the business has such difficulty understanding why the team would want to make a change to the software development process. When the business lacks the insight into how a process change is implemented, their appreciation for the hard work behind the change may be lacking, and so they do not understand that the entire organization will benefit from the newly implemented changes, even if they were not consulted on areas identified as needing improvement.

It is not that the business necessarily sees the development team as lazy. On the contrary, I think the notion is that software developers are heads-down in code all day. However, they are not aware of the learning and dedication that goes into writing software. From their viewpoint, the software development process is vanilla. Requirements are produced, code is written, software is shipped, and the client is (hopefully) happy. They are unaware of all the steps in between required to produce the desired bit of software, and so are unaware of the needs for and advantages of changing this process.

It is therefore important to keep the business informed of wins as they occur, when implementing process changes. I mentioned back in Chapter 5: Working with Key Players that, as a process is undergoing change, whether the business is aware of the impact that it will have or not, you should keep them in the loop about any sort of hurdles that you may be running into. The same is true of any wins. Regular check-ins with key players within your organization will alleviate a lot of the tension and blowback that could occur as the result of obstacles along the way. A sure-fire way to lose confidence is to experience an issue that everyone believed would never be one.

To capitalize on the wins, I suggest you produce a series of status reports for the business as the new process begins to positively impact the ongoing project. These reports can help to generate buzz around the process. If you or the business are not interested in status reports, then you can invite interested coworkers to your daily standups. You may find that keeping even just one person in the organization aware of the positives can positively impact the dissemination of information to the remainder of the business. At one company I worked for, we involved a couple of our customer service staff in our daily stand-up meetings. They would stand silently on the side and just listen in. After a few weeks, we found out that they had been relaying the information back to their team, and that their team was more prepared for events that were taking place as a result.

Bear in mind that it is also important to keep the business informed of the failures your team has been able to overcome. Let them know what happened and how it was remedied. We spoke about team retrospective meetings in Chapter 7: Winning (and Losing) as a Team. I have been in larger organizations that corral members from every department and have a postmortem meeting to discuss major incidents that occurred. The point of these meetings is to understand what problem occurred, then establish what went right and what went wrong while coming up with a remedy. This mindset encourages healthy conversation surrounding issues that arise and aside from giving the team the opportunity to capitalize on their wins by showcasing what they did to correct a major issue, it also gives the team the opportunity to highlight what they learned from their mistakes. Never burden the business with every detail of what is going on with the team, but make sure that they are aware that the team is capable, and they are doing what it takes to keep the client happy.

If there is tension between the software development team and the remaining business in your organization, you would likely benefit from modernizing your process more slowly, and in smaller chunks; smaller even than those we have been discussing so far. This will allow your team to continue working on existing features that have been scoped out while you update your process for new features and requests. However, do not lose sight when something new comes on; start to think about how you could break your newer tasks down smaller.

One of the positive effects of a slow-roll process change is that the business gets to see the old process versus the new process, side-by-side, with the added benefit that they could possibly help in identifying potential roadblocks. Your team can demonstrate their ability to readjust quicker than they could in the legacy process and the executive-level members of your organization can be involved in that new process.

Executive-level members are likely more interested in the money that can be saved or made by the software development process than in discussions of transparency or the like, so for them the biggest win that you can show is that the implementation change occurred without becoming a costly mistake. You can showcase the efficiency of the process by comparing two similar cases; for example, you might present a case wherein a client had requested changes after development had finished, which is usually costly and time consuming. In the new process, the level of transparency that came with implementing bite-sized wins created a more-frequent flow of deliveries, which allowed the client to identify their need for a change in requirements much earlier in the process. This way, a software developer simply needed to check in a small code change and the requested change was implemented right away. The amount of work that is required for changing requirements in the middle of the software development cycle is far less than the amount of work required when changing the requirements after a full product has been delivered. The business and the client will appreciate the cost savings. This holds especially true for the hands-on client, who is more likely to change their requests after seeing what has been delivered. These clients are not necessarily looking for reasons to complain, but they tend to find fault in many deliveries.

The business will also appreciate any updates you can give on the client's happiness with the process. Always include the business in your celebrations of wins so that they can understand the hard work that went into it and understand that you are keeping the clients' best interests in mind.

Client visibility

Approval from your client is going to be the greatest value-add for your new process change. Once your team has finished implementing the bite-sized win mindset, your process will inevitably increase visibility for your clients. The client's ability to see updates more frequently will be an excellent way to ensure a positive reception. One of the main goals of the bite-sized approach is to provide the client with a way to self-report on the status of projects. When they have frequent deliveries with more fruitful information, they will be able to find answers for themselves.

As previously stated, there are different ways to handle the hands-on client versus the hands-off client, but the ways that you can capitalize on their positive experiences are similar. For a client who is hands-off and not very aware of your development process, you might have weekly check-ins to ensure the project is on-track. Perhaps the client presents ideas or raises concerns about previous deliveries during these meetings. After adopting the bite-sized approach, you may start to notice that these meetings contain more valuable information because the client has more to review beforehand thanks to the new cadence of delivery. One of the best indicators that your client has found value in the bite-sized delivery process is that they will be in contact with your team before scheduled meetings. This is typically a good sign because it means they are no longer waiting for completed projects before they ask questions or raise concerns. Now, they have enough information available during the software development process to raise concerns on a frequent basis. Clients enjoy scheduled touchpoints since it gives them the opportunity to get a feel for how the project is running, but when they feel empowered to break away from that schedule and make ad-hoc requests, they are also starting to benefit from the bite-sized wins that your team has put in place.

The hands-off client giving accolades directly is generally one of the best signs that your process is working, since they likely never knew that your process changed in the first place. The hands-on client, on the other hand, may make a remark or two about how they have noticed more software deliveries and, whether they realize it yet or not, that will be the point at which they show that they have started to benefit from the bite-sized win mindset.

To ensure you are capitalizing fully on your wins with your client, I suggest being honest about the amount of work that has been completed and, if possible, showing them regular metrics related to delivery schedules. If the client sees a report on work delivered on a weekly basis in the old process versus how much is now being delivered on a weekly basis, they will understand that some change in the backend has benefited them in a positive way.

When I was working for the internet marketing firm, we would handle all aspects of web-based marketing for our clients, and they were not required to be hands-on. All that was required of them was to sit back and enjoy the fruits of our hard labor. My manager was a very business-savvy lady who understood that no matter how hard you tried, there would always be someone rooting against you, so she took pre-emptive precautions to ensure that we did not run into issues with a rogue staffer trying to sabotage our hard work. One way she handled this was to host quarterly presentations in which she would talk about our team's successes and point out how we overcame failures and what we did to prevent similar failures in the future. Even though we expected our clients to be hands-off, they appreciated the presentations and felt secure that their company had been left in the correct hands. This level of insight into our process gave our clients a greater sense of trust in our team's ability to operate without direct oversight. Our clients had enough information to know which struggles our teams faced, what decisions were made, why we made them and how we and they benefitted from the direction that we chose. Without our manager's constant updates, it is possible that the client would have questioned our value.

There are a lot of tools and methodologies available for tracking your team's work to help you present it to the client. We might not all be savvy in presenting our own victories, so it may take some third-party tools to help track successes over time.

One good metric to show off is a burndown chart. The point of the burndown chart is to show that work is steadily delivered. Therefore, highlight that, before you started working with smaller tasks, it is possible that you were delivering inconsistently, whether due to tasks being too large or because your team faced frequent roadblocks due to incomplete requirements.

Bite-sized tasks allow your team to deliver work in a shorter amount of time, thus showing a better burndown during their sprint.

Another common aspect to highlight with a burndown chart is that, in a traditional workflow, software developers take on a lot of work at the beginning of the sprint, do not start committing work until about midway through the sprint and, by the end, often still have outstanding work, which explains why the burndown chart never hits zero. Switching to the bite-sized win mindset within your team, you can see work drop off at a more frequent and natural pace.

When you can show the client that your team is succeeding in your sprints, they will understand that something happened in the backend that means they are getting their money's worth out of the project. The client generally has an inherent anxiety throughout the process that not enough work is being completed, and then, at the very end of the project, everything will suddenly be delivered at one time. This can result in months of anticipation and anxiety due to lack of visible results. In order to capitalize on your team's wins, you want the client to see that frequent deliveries give them a level of transparency that they can leverage to ensure their project solves the problem that they originally had no solution for.

Summary

When implementing the bite-sized win mindset, it will be important to consider how you will prove to everyone involved in your software development process that the new process is valuable. You may find that changing the process is initially a damper on team morale since they are used to working with a process that they found to be tried-and-true. The business may think your team is wasting time and money since they just want to see work being delivered. Your client may not even notice that a change has happened and could be missing an opportunity to benefit from frequent deliveries. You should find ways to constantly highlight the benefits of your team's new process and ensure that your key players are satisfied with the new direction that your team is headed. We have one shot at a first impression, so make it a good one!

I ALWAYS RECOMMEND TRYING TO FIND A WAY TO **CAPITALIZE ON YOUR BITE-SIZED WINS**

Chapter 9 | The Obligatory Wrap-Up Summary

My career as a software developer began around 2006. In my time, I have seen many iterations of team organization and many interpretations of the software development lifecycle. When I began, my company had us using software development tools like Dreamweaver because it was basically the Wild, Wild West and we all liked to watch the world burn. These days, we know better. Yet one thing that has not changed over the years is a team's desire to succeed. I have seen teams go through many iterations of methods for requirements gathering, ticket creation and business buy-in.

I strongly believe in striving towards a bite-sized win mindset within your organization. This is the single most important concept that I continue to teach my teams today. Even before I had this practice properly formulated with a fancy-schmancy term, I was a firm believer in delivering smaller in order to make a bigger impact. Visibility into the development process is key, and easily achieved with the bite-sized win approach. Quality will always be important, but lack of transparency can kill a project.

I have worked with developers and managers in the past who believe that there will always be a rift between the development team and the business, but I reject this as an acceptable school of thought. In fact, I have heard the opinion that "singing 'Kumbaya' with the business will always lead to failure." I do not accept that. I have mentioned that all entities involved in the software development process are concerned with one thing: ~~keeping their jobs~~ Customer Satisfaction.

One thing that you cannot teach to anyone in an organization is how to take pride in their work. A few months ago, I felt like trolling and posted a simple question on LinkedIn: "For those of you who did not pay attention at your company's orientation, how are you doing in your current role?" Nobody responded. The lack of response may be an indication that nobody got the joke ... or maybe it was because everyone knows that company orientation is often not taken seriously. I, however, believe that the company orientation is imperative because it starts you out in the same mindset as everyone else in the organization. Undoubtedly, you will not later recall the name of the handsome fella who founded the organization in 1826, but you will gain an appreciation for the company's roots. I believe that pride in your work begins with pride in your company. We want to feel as though we are working towards something greater, and if we lack faith in our employer, we will inevitably feel that we are just working for the sake of work, and not sharing in the company's vision.

Now, I realize that you have hung onto every word that I have said in this book, will never forget a single lesson that was given, and will run this advice over to your team tomorrow morning and instantly start making positive changes to your software development process. However, for *those of us* who were not as diligent about paying attention, here is a quick recap on the most important takeaways from each chapter.

Painstakingly summarizing every chapter

Chapter 1: What is a Bite-Sized Win? The important thing to take away from this chapter is that a bite-sized win is not specific to software development. It is a concept that can be used in your everyday life. The concept itself stems from our natural reactions to frequent reassurance: we like to know that what we are doing is correct and we like to hear it often. When you separate work into smaller, bite-sized pieces, you will notice that the software development team, clients and the business will all start to appreciate the work that is being done even before it is ready to ship.

Chapter 2: How Teams Self-Organize. As a leader, it is important to understand when your team is ready to work independently. Different Agile methodologies, such as Scrum, aim to give the power back to the team and let the team manage the way that their work is handled. If you are an active member of the team, you do not want to stick out as the leader but should rather be a *contributing* member. The team must be willing and able to take care of the delegation and completion of work in order to ensure the team's undeniable success.

Chapter 3: Tasking for the Bite. You and your team will need to formulate rules around how tasks will be written using a bite-sized win mindset. I would be willing to bet that your team's current method for story-creation is to write a story that focuses on a chunk of the big picture, assign it to one software developer who possibly wrote some subtasks that will only be handled by themselves, and then accept that you will not likely see the software developer again until that story has been completed. There is an obvious lack of feedback for the client and the business in this structure, and nobody knows the true status of the task until the software developer is willing and able to present new information. The solution is the bite-sized task: the development team will need to separate the requirement into tasks in such a manner that each task delivers enough to show real progress without asking for so much that it negatively impacts day-to-day working activities.

Chapter 4: The Toss 5 Method. This chapter demonstrated a way of helping your team to think about how they can achieve goals or organize work in a bite-sized manner, in both their daily life and work life. I came upon the idea behind the Toss 5 Method when I was researching minimalism: toss five things a day (over a certain time period) to declutter your life. I like to use it for myself and my software teams to get into the mindset of challenging yourself to take on smaller tasks that, over time, build up to a completed project. In the example provided in this chapter, you toss five items every day from your office, your home, or even your car, to clean up your surroundings. By committing to a strict schedule of throwing away five items every day, within a month, you will have thrown away roughly 150 items that were just clutter in your life. You can then relate this method back to code development: apply this to ticket creation or code refactoring or learning something new, and you can clean up any low-hanging work items and slowly transform the team's way of organizing. The sky is the limit on this one, but the main goal here is to ensure that your team can learn how bite-sized wins improve the workflow over time.

Chapter 5: Working with Key Players. In Chapter 5, we get into the meat and potatoes of how you should approach different types of people involved in the software development process as you implement the bite-sized win approach. This is not an easy road, but it is an important one. We break those who have an interest in the project's success down into three main categories: clients, the business and software developers. Each category has different concerns, and when you change the software development process, it is important to deal with everyone in the appropriate manner. One of the key emphases from this chapter is to ensure that you provide regular updates to everyone throughout the process. If any one of the players in your project feels as though they have been kept in the dark, it can have a ripple effect that negatively impacts everybody throughout the team. Try your best to give as much information as is appropriate; gauge whether that information is enough to keep everyone happy and excited about the process and the changes to come.

Chapter 6: Separation of Concerns. The concept of separation of concerns emphasizes the importance of keeping work separated, based on logical grouping. It is possible that your software developers are already utilizing this concept, segmenting code and project architecture into logical categories and processes so that the affected areas (such as a method in code) do not overlap with logic that belongs to a different area. We can extend this idea to general team practices. For example, you can break up larger tickets into smaller tasks based on different *concerns*, where a concern is a single function that has an input and an expected output. When breaking down larger tasks into bite-sized tasks, you want to ensure that the delivered unit of work is small enough to be delivered quickly, but not so small that it does not achieve any goal or progress the project. Take the time to examine how your tickets depend on each other. Ensure that your developers know that when one ticket is completed, it will allow the next ticket to start development. If tickets can be worked on simultaneously, write them in that manner. Do not, however, be afraid to delay the delivery of certain tickets based on the completion of two or more dependency tickets. The point of this process is to separate the work as small as you can and create a way to deliver on an iterative schedule.

Chapter 7: Winning (and Losing) as a Team. Celebrating your wins is easy. We do it every day. What we are never taught is how to celebrate our failures. Failures can provide learning opportunities that cannot be attained from a win.

Of course, it is not always appropriate to celebrate failure. For example, if your team has repeated a failure on the same topic, it means that the team did not learn a lesson and there is nothing to celebrate.

Remember to share your team's wins with your key players. I always tell my teams that we win and lose as a team, regardless of who did what. If an individual contributed significantly to a success, that team member deserves recognition. However, the win did not occur in a vacuum, so the whole team should be recognized for helping to support that member throughout the win. On the other hand, when a failure occurs, everyone may be quick to point fingers, but the team must remain humble and likewise share in the failure. Remember the adage, "There is no 'I' in team": there may be times when you feel like you are the only one doing the real work, but the reality is that that no matter what you are doing, even if you are working alone in a bubble, you did not do everything alone. If you were the one who found the issue and decided to fix it quietly, it may be that a conversation with a team member helped you understand the cause. Even if you never spoke with anyone about the issue, someone had to be there to check the work, and someone also had to be there to deliver the work to the server or write the process that automatically uploaded the work. Therefore, I reiterate that you should always make sure that the team understands that they win and lose together. In the end, they are going to understand the value of each other's position within the team.

Chapter 8: Capitalizing on the Wins. In the penultimate chapter, I discuss how we can reap further benefits of bite-sized wins. By ensuring you give adequate recognition to the work that your team has delivered, you have boosted the morale of everybody involved, as well as their confidence in the bite-sized win system. It is very possible that when you start to implement a new software development process, you see hesitation and doubt in various members of the organizations. If people are not receiving frequent reassurances, they are not going to fully understand the benefits of an improved system. Ensure that your team's wins are highlighted to those outside the core team, and make sure that they continue to buy in to the new process changes. Do not let your successes go unnoticed; recognize and properly capitalize on the wins that occur.

Afterword

I wanted to take a moment to thank you, the reader, for reading this book. This book has been a labor of love for me and is a culmination of the hard work that I have put in throughout my career. Understanding teams and helping them thrive is truly a passion of mine and I hope that I can, in some way, help your team organize in a way that makes sense for them. Regardless of my passion for the bite-sized win mindset, I still recognize that this process is not a one-size-fits-all for teams all around the world, but I believe that it can be adapted and adopted to have massive benefits for many teams worldwide.

When considering changes to your software development process, take the time to examine all avenues available to you and check in with your team regularly to get a feel for how they are getting on with the changes. If a process does not work, do not be afraid to change it. If you feel passionate about a new method of organization but you just cannot seem to get it right, do not be afraid to admit that it is not the right fit for your team and rule it out. It takes a lot of time and effort to get your team working like a well-oiled machine and, as time goes on, the oil will run out. Processes were meant to be changed and as we innovate, we find new ways to do the work that we feel so passionate about.

From the depths of my heart, thank you for reading!

About the Author

Derek Torrence was born and raised in sunny South Florida, where he received all his education. Originally planning on teaching, Derek began studying English at the local community college, but quickly switched to Information Technology and, ultimately, software development. Throughout his tenure as a software developer, Derek has worked for small businesses as well as corporate-level organizations. His love for organizing teams has followed him throughout this time.

Derek has since left South Florida and moved to the Midwest, where he has found many similarities in the way that teams organize and deliver software. He is still a software developer, but has since started to specialize in software delivery processes. His passion for organizing teams continues to reign supreme and he hopes to continue his tradition of influencing software developers all around the world to adopt more of a Bite-Sized Win Mindset.

www.ingramcontent.com/pod-product-compliance
Lightning Source LLC
Chambersburg PA
CBHW061021220326
41597CB00016BB/2027